Stability of Free Surfaces in Single-Ph.

Two-Phase Open Capillary Channel Flow in a

Microgravity Environment

I0037947

# Stability of Free Surfaces in Single-Phase and Two-Phase Open Capillary Channel Flow in a Microgravity Environment

Vom Fachbereich Produktionstechnik

der

UNIVERSITÄT BREMEN

zur Erlangung des Grades

Doktor-Ingenieur

genehmigte

Dissertation

von

Peter Johnathan Canfield, M.Sc.

Gutachter: Prof. Dr.-Ing. habil. Michael E. Dreyer
Prof. Dr.-Ing. habil. Udo Fritsching

Tag der mündlichen Prüfung: 27.02.2017

**Bibliografische Information der Deutschen Nationalbibliothek**

Die Deutsche Nationalbibliothek verzeichnet diese Publikation in der
Deutschen Nationalbibliografie; detaillierte bibliografische Daten sind im Internet
über http://dnb.d-nb.de abrufbar.

1. Aufl. - Göttingen: Cuvillier, 2018

Zugl.: Bremen, Univ., Diss., 2017

© CUVILLIER VERLAG, Göttingen 2018

Nonnenstieg 8, 37075 Göttingen

Telefon: 0551-54724-0

Telefax: 0551-54724-21

www.cuvillier.de

Alle Rechte vorbehalten. Ohne ausdrückliche Genehmigung des Verlages ist
es nicht gestattet, das Buch oder Teile daraus auf fotomechanischem Weg
(Fotokopie, Mikrokopie) zu vervielfältigen.

1. Auflage, 2018

Gedruckt auf umweltfreundlichem, säurefreiem Papier aus nachhaltiger Forstwirtschaft.

ISBN 978-3-7369-9726-4

eISBN 978-3-7369-8726-5

## Abstract

Low Bond number open capillary channel flows have been shown to exhibit collapsing free surfaces when a critical flow rate is exceeded, a phenomenon that is referred to as choking. As shown in this work, the critical flow rate can be pre-determined with sufficient accuracy for the presented channel geometry when certain boundary conditions are known *a priori*. The presented model that describes the flow rate limitation of stable liquid flow through the open channel is examined and compared to numerical simulations and experimental studies. In addition, the characteristics of the supercritical domain, in which bubbles are ingested passively into the flow in the channel, are described and a new model for bubble formation via choking is proposed.

## Zusammenfassung

Offene Kapillarströmungen im Regime niedriger Bond-Zahlen weisen kollabierende Flüssigkeitsoberflächen auf, wenn ein kritischer Volumenstrom überschritten wird; ein Phänomen, das Choking genannt wird. In dieser Arbeit wird gezeigt, dass dieser kritischer Volumenstrom für die vorliegende Kanalgeometrie mit ausreichender Genauigkeit vorhergesagt werden kann, wenn bestimmte Randbedingungen des Strömungsproblems *a priori* bekannt sind. Das vorgestellte Modell zur Beschreibung der Volumenstrombegrenzung in einem offenen Kapillarkanal wird untersucht und mit numerischen und experimentellen Studien verglichen. Außerdem werden die Charakteristika des überkritischen Regimes, in das Gasblasen in die Strömung eingesogen werden, beschrieben und ein neues Modell für den Blasenentstehungsprozess im Choking vorgestellt.

# Contents

# List of Figures

# List of Tables

# List of Symbols

## Variables and constants

| Symbol | Units | Description |
|---|---|---|
| $\alpha$ | ° | Half-angle of vertex in channel's cross-section |
| $\beta$ | | Volume fraction |
| $\gamma$ | ° | Apparent static contact angle |
| $\gamma_{eff}$ | ° | Effective contact angle |
| $\Gamma$ | | Geometrical factor |
| $\lambda$ | m | Wave length |
| $\Lambda$ | | Aspect ratio |
| $\varphi$ | ° | Half-angle of circular segment described by interface in the $y'z'$-plane |
| $\phi$ | | Void fraction |
| $\Psi$ | | Geometrical factor |
| $\tau'_{i,j}$ | $N\,m^{-2}$ | Deviatoric stress tensor |
| $\theta$ | ° | Angle of inclination of the interface's normal vector |
| $a$ | m | Half channel width (along $y'$-axis) |
| $A'$ | $m^2$ | Area of cross-section |
| $A'_0$ | $m^2$ | Area of cross-section at $x' = 0$ |
| $b$ | m | Channel height (along $z'$-axis) |
| $C_0, C_1$ | | Fit coefficients |

| | | |
|---|---|---|
| $d'$ | m | Distance between channel height and the inter-face for $y' = 0$ within a cross-section |
| $d'_h$ | m | Hydraulic diameter |
| $D_{Ayy}$ | m | Ayyaswamy hydraulic diameter |
| $f'$ | s$^{-1}$ | Frequency |
| $f'_{C2}$ | s$^{-1}$ | Operation frequency of valve C2 |
| $F'_\sigma$ | N | Volumetric surface tension force |
| $F'_A(k')$ | m$^2$ | Function of cross-section area of streamtube |
| $g'_i$ | m s$^{-2}$ | Gravitational acceleration |
| $h'$ | m$^{-1}$ | Mean curvature $(2H' = h')$ |
| $H'$ | m$^{-1}$ | Mean curvature |
| $k'$ | m | Height of interface in $x'z'$-plane |
| $k'_c$ | m | Critical contour height |
| $K_0$ | | Coefficient for the static term of the dimension-less pressure boundary condition |
| $K_1$ | | Coefficient for the friction term of the dimension-less pressure boundary condition |
| $K_2$ | | Coefficient for the convective term of the dimen-sionless pressure boundary condition |
| $K_f$ | | Friction factor |
| $l'$ | m | Channel length (along $x$-axis) |
| $\tilde{l}$ | | Dimensionless channel length |
| $l'_e$ | m | Entrance length |
| $L'$ | m | Characteristic length scale |
| $m'$ | kg | Mass |
| $N_{cells}$ | | Number of cells |
| $p'_0$ | N m$^{-2}$ | Pressure at x=0 |
| $p'_a$ | N m$^{-2}$ | Ambient pressure |
| $p_c$ | N m$^{-2}$ | Characteristic pressure |
| $p'_l$ | N m$^{-2}$ | Pressure in liquid |
| $p'_n$ | N m$^{-2}$ | Irreversible pressure loss in entrance region |

| $\Delta p'_{cap}$ | $\mathrm{N\,m^{-2}}$ | Capillary pressure difference |
| $P'$ | m | Wetted perimeter |
| $Q'$ | $\mathrm{m^3\,s^{-1}}$ | Flow rate |
| $Q'_{crit}$ | $\mathrm{m^3\,s^{-1}}$ | Critical flow rate |
| $Q'_{FM}$ | $\mathrm{m^3\,s^{-1}}$ | Measured flow rate |
| $Q'_P$ | $\mathrm{m^3\,s^{-1}}$ | Flow rate setting for the pump |
| $Q'_G$ | $\mathrm{m^3\,s^{-1}}$ | Time-averaged gas flow rate |
| $r'_W$ | m | Wetted length of side wall in a cross-section |
| $R'$ | m | Radius of circular segment described by interface in the $y'z'$-plane |
| $R'_1, R'_2$ | m | First and second principal radii of curvature |
| $R'_{bi}$ | m | Radius of injected bubbles |
| $t'$ | s | Time |
| $\Delta t'_{dc}$ | s | Duty cycle of valve C2 |
| $T'$ | K | Temperature |
| $u'_{i,\beta}$ | $\mathrm{m\,s^{-1}}$ | Slip velocity |
| $u'_i$ | $\mathrm{m\,s^{-1}}$ | Volumetric flux |
| $v'$ | $\mathrm{m\,s^{-1}}$ | Velocity |
| $v'_{Ayy}$ | $\mathrm{m\,s^{-1}}$ | Ayyaswamy velocity |
| $v_c$ | $\mathrm{m\,s^{-1}}$ | Characteristic velocity |
| $V'_b$ | $\mathrm{m^3}$ | Volume of ingested bubbles |
| $V'_{bi}$ | $\mathrm{m^3}$ | Volume of injected bubbles |
| $w'_f$ | $\mathrm{N\,m^{-2}}$ | Viscous stress term in one-dimensional momentum equation |
| $x', y', z'$ | m | Cartesian coordinates |

# Physical properties

| Symbol | Units | Description |
| --- | --- | --- |
| $\mu$ | $\mathrm{N\,s\,m^{-2}}$ | Dynamic viscosity |
| $\nu$ | $\mathrm{m^2\,s^{-1}}$ | Kinematic viscosity |
| $\rho$ | $\mathrm{kg\,m^{-3}}$ | Density |
| $\sigma$ | $\mathrm{N\,m^{-1}}$ | Surface tension |

# Subscripts

| Index | Description |
|-------|-------------|
| 0,1 | Indicating that a dimensionless variable is located at the inlet or outlet of the test channel respectively |
| $1P, 2P$ | Indices pertaining to single-phase or two-phase flow respectively |
| $1D$ | Index pertaining to numerical computations using the one-dimensional model |
| $3D$ | Index pertaining to three-dimensional numerical simulations |
| $b$ | Index pertaining to a gas bubble |
| $C2$ | Index pertaining to gas injection via valve C2 |
| $ing$ | Index pertaining to gas ingestion |
| $exp$ | Indices pertaining to experiments |
| $FM$ | Index pertaining to the flow meter |
| $G, L, S$ | Indices pertaining to gaseous, liquid, or solid phase |
| $i, j, k$ | Tensor dimensions |
| $in$ | Index pertaining to the test channel's inlet |
| $m$ | Index pertaining to an averaged mixture property |
| $out$ | Index pertaining to the test channel's outlet |
| $P$ | Index pertaining to the pump |
| $sc$ | Index pertaining to supercritical flow |

# Dimensionless numbers

| Symbol | Description |
| --- | --- |
| Bo | Bond number |
| Oh | Ohnesorge number |
| Re | Reynolds number |
| Su | Suratman number |
| We | Weber number |

# Abbreviations and Acronyms

| Abbreviation | Description |
| --- | --- |
| $\mu g$ | Microgravity |
| BB | Bubble |
| BI | Bubble Injector |
| C1, ..., C11 | Valves |
| CCF | Capillary Channel Flow |
| CFD | Computational Fluid Dynamics |
| CT | Compensation Tube |
| ESS | Electrical Subsystem |
| EU | Experiment Unit |
| FPC | Flow Preparation Chamber |
| GR | Channel geometry: Groove |
| HSHRC | High Speed High Resolution Camera |
| ISS | International Space Station |
| K1,K2,K3 | Plungers |
| MSG | Microgravity Science Glovebox |
| ODU | Optical Diagnostics Unit |
| PP | Channel geometry: Parallel Plates |
| PSC | Phase Separation Chamber |

| | |
|---|---|
| TC | Test Channel |
| TU | Test Unit |
| VE | Vertex |
| WE | Channel geometry: Wedge |

# Chapter 1

# Introduction

Open capillary channels are structures that contain a liquid with one or more free surfaces (gas-liquid interfaces) and in which capillary forces dominate the characteristics of the flow rather than gravitational forces. This can be the case in environments where the influence of gravity is reduced in comparison to other forces or compensated, e.g. in space, but also occurs on Earth when the characteristic length scale of the flow is small ($\approx 10^{-3}$ m). The increased importance of capillary forces in a reduced gravity environment opens the door for passive transport and control mechanisms that rely on surface tension, wettability, and geometry to perform tasks that would otherwise require bulkier and heavier equipment or less reliable techniques involving movable parts [63]. For example, open capillary channels are used in propellant management devices (PMDs) of space vehicles to position and transport liquids within surface tension tanks. During acceleration phases the liquid bulk orients itself due to body forces acting on its mass, but during non-acceleration phases the liquid may distribute itself freely within the tank. Consequentially, the outlet of the tank may lose contact with the liquid rendering the remaining propellant in the tank inaccessible for further operations unless the liquid can be transported towards the outlet again. One of the purposes of PMDs is to prevent the liquid from becoming inaccessible. Surface tension tanks employ vanes, which are narrow and thin sheets of metal that may be positioned parallel

or perpendicular to the tank's wall to form open capillary channels [31]. Using
vanes to transport liquid from a bulk reservoir within the tank towards its outlet
port is weight-efficient and increases the device's reliability. Similar benefits can be
of importance in other liquid managements systems such as those used in a space
vehicle's life support systems or storage tanks. Capillary channels are also used
for managing liquids in micro-electromechanical systems for lab-on-a-chip devices
[39, 69], effectively reducing the need for valves and pumps in micro-scale devices
that are utilised in the biological and chemical industries.

However, previous studies [22, 31, 46] have shown that free surfaces can collapse
and cause gas ingestion in open capillary channel flows when a critical, maximum
flow rate, $Q'_{crit}$, is exceeded[1] (compare figure 1.1). This flow phenomenon is referred
to as 'choking'. In reduced or compensated gravity, open capillary channel flows
are subjected to changes in cross-sectional area due to pressure loss or gain in flow
direction. The free surface of the open channel behaves like a flexible wall and the
pressure difference between the liquid in the channel $p'_l$ and the ambient pressure
$p'_a$ are balanced by the local capillary pressure difference induced by surface tension
and the local mean curvature of the free surface. Viscous and convective pressure
losses within the channel increase with the flow rate and lead to a higher pressure
difference across the interface which in turn is balanced by an increase of curvature.
If the flow rate is increased further, at some point the maximum mean curvature
of the free surface is no longer sufficient to balance the pressure difference and gas
ingestion occurs at the gas-liquid interface; the flow in the channel is choked. The
critical flow rate is defined as the maximum flow rate before choking occurs.

In some applications, choking may be detrimental to the flow system due to the
ingested gas bubbles. Alternatively, bubble generation may be desired in other flow
systems and achieved passively in this way. Understanding the mechanisms of open
capillary channel flow will help optimise designs that are now in use and may widen

---

[1]With regard to notation, symbols for dimensional variables are distinguished by a prime (')
following the letter (e.g. $Q'$). Constants and dimensionless numbers or dimensionless variables are
written without primes (e.g. $Q$).

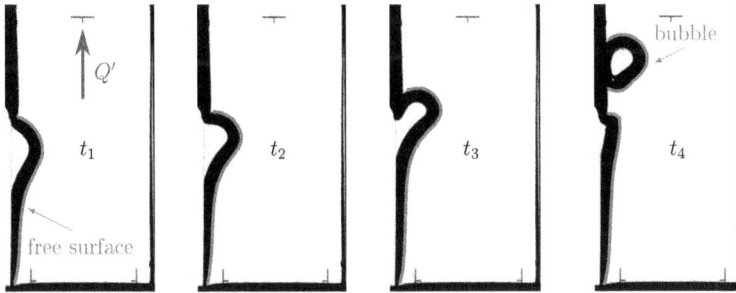

Figure 1.1: Time sequence of choked flow in an open capillary channel. Side view with flow from bottom to top. As time progresses from $t_1$ to $t_1$, the interface bends into the channel and a gas bubble is ingested.

the field of viable applications. It may be noted at this point that the presented capillary channel flow experiment is an ideal case with a simplified geometry, well-defined boundary conditions, and no residual acceleration (as far as can be provided on board the ISS). The actual situation in liquid management systems on a space vehicle may differ widely in terms of residual acceleration, geometry, and boundary conditions.

## 1.1   Scope

This work is concerned with the execution and analysis of experiments on flow rate limitation in isothermal, forced, single-phase and two-phase bubbly open capillary channel flow in a microgravity environment. It should be noted that forced capillary flow is considered here (i.e. via a pressure gradient that is generated by a pump) in contrast to capillary driven flow, in which the driving force of the flow is the capillary pressure flow of a propagating meniscus. The investigated test channel is displayed in figure 3.3. The channel's cross-section is an isosceles triangle and constant over its length $l'$. The triangular cross-section is defined by its height $b$ and the opening angle $2\alpha$ located in its vertex. The channel's base width $a$ is defined as $a = 2b \tan \alpha$.

The test channel is open on one side along $l'$ and is surrounded by a static gaseous atmosphere with a given pressure $p'_a$ which is higher than $p'_0$, the pressure in the liquid at the inlet of the test channel ($x' = 0$). The liquid within the channel is fully wetting with a static contact angle of zero ($\gamma = 0°$) and is saturated with the surrounding gas to prevent diffusion. The flow rate $Q'$ is constant for each experiment and set at the outlet of the channel, the pressure at the inlet is well-defined and assumed to be known. Under these conditions, the pressure difference between the gaseous atmosphere and the flowing liquid in the open channel is balanced by the curvature of the free surface in accordance with the Young-Laplace equation [12], in which the product of surface tension and curvature define the capillary pressure

$$\Delta p'_{cap} = -\sigma \left( \frac{1}{R'_1} + \frac{1}{R'_2} \right), \tag{1.1}$$

where $\sigma$ is the surface tension and $R'_1$ and $R'_2$ are the first and second principal radii of interface curvature, respectively. The interface is assumed to be symmetrical with respect to the $x'z'$-plane. Both $Q'$ and $l'$ are varied to determine the critical flow rate $Q'_{crit}$ as a function of $l'$. Furthermore, a gas injection device is located at the inlet of the test channel which is used to observe the influence of bubbly, two-phase flow on the channel's flow rate limitation.

The gathered data is compared with numerical results that are produced with a previously published one-dimensional analytical model and with three-dimensional numerical simulations. In addition, the analytical model is modified to account for two-phase bubbly flow and a new model describing bubble formation in the supercritical regime is postulated. The intent of this thesis is to compare experiment results with those achieved through computational means with the intent of model verification. An overview of the experimental and numerical data that was evaluated to this effect is presented in table 1.1.

This thesis is structured in the following manner: First, an overview of past research in the relevant fields field of fluid mechanics is presented followed by a short introduction into surface tension and an overview of the various multiphase flow models. In chapter 2, the one-dimensional model for open capillary channel

Table 1.1: Overview of the presented data. 1P and 2P represent single-phase and (injected) two-phase flow respectively. Two-phase flow is understood here only as forced gas injection into the test channel in contrast to gas bubble ingestion that always occurs in supercritical flow. Flow is understood to be critical when the upper flow rate limit is reached.

|  | 1P 2P | 1P 2P | 1P 2P |
|  | subcritical flow | critical flow | supercritical flow |
| --- | --- | --- | --- |
| Experiment | yes no | yes yes | yes no |
| 3D CFD | yes no | yes no | no no |
| 1D model | yes no | yes yes | yes no |

flow through an isosceles triangular wedge is presented. In the following chapters, materials and methods used to gather data in experiments and numerical simulations are described and all results are discussed and compared in chapter 6. Also, the supercritical flow domain is described and a new bubble formation model is presented and compared to experiment results in a rectangular capillary channel in chapter 7.

# 1.2 State of the Art

## 1.2.1 Open Capillary Channel Flow

Initial analytical work on the performance of vanes as propellant management devices was conducted by Jaekle [31]. In this paper, he demonstrates the aspect of flow rate limitation in open capillary channel flow through propellant vanes in a surface tension tank and compares the phenomenon to flow rate limitation in compressible flow within closed converging-diverging ducts or incompressible flow through closed ducts with elastic walls. All these duct flows have in common a non-zero $dA'/dx'$ term (or $d\rho/dx'$ in compressible flow) in the respective equations for mass conservation due to the non-constant nature of the flow path's cross-section or the density of the gas, respectively. Flow rate limitation in compressible flow is based on the speed of sound, or the speed of density waves. Likewise, in open capillary channel flows and flow through elastic tubes, flow rate limitation, or choking, is defined by the velocity of a longitudinal capillary wave, or put otherwise, the speed of area waves. Jaekle uses a one-dimensional model to predict flow rate limitations for T-shaped vanes assuming steady, fully developed laminar flow without accounting for the effects of area variation on the viscous term.

In similar fashion, the work of Shapiro [53] is noteworthy due to the presentation of an analytical model for flow rate limitation in elastic, collapsible tubes. Shapiro defines a Speed Index for incompressible flow through elastic tubes as the ratio between local flow velocity and long wavelength phase velocity. Choking in an elastic tube occurs when the Speed Index ratio reaches unity. Ultimately, both Shapiro and Jaekle use models to show that wave speed-based flow rate limitation, that is traditionally associated with compressible flow such as in Laval nozzles, can indeed occur in incompressible flow when channel area variations are possible.

Experiments on flow rate limitation in a capillary channel composed of parallel-plates were conducted by Dreyer et al. [22] and by Rosendahl et al. [45] using a drop tower facility to mitigate hydrostatic effects. Their findings substantiated the hypothesis that open capillary flows are subject to a maximum steady volume flux, or a flow rate limitation, beyond which gas ingestion occurs across the free surface. Additional experiments were performed on sounding rocket flights to increase the duration of the experiments from a few seconds to several minutes which allowed further observation of the supercritical regime in which periodic gas ingestion occurs. Rosendahl et al. [46] compared the experiment results of drop tower experiments and an experiment onboard sounding rocket TEXUS-37 with an improved one-dimensional model that incorporated both principal radii of curvature of the free surface. The numerical predictions for the profiles of the free surfaces were found to be in good agreement with experimental findings. In addition, it was shown that the Speed Index ratio as defined in [31] and [53] is indeed verified by experiment results.

Further experiments were conducted on sounding rockets TEXUS-41 and TEXUS-42 to determine the effect of channel length on the stability of the free surface [44, 47]. It was found that increasing the channel length at a fixed flow rate could also lead to choking in the capillary channel. The extension of the one-dimensional mathematical model to incorporate accelerated flows was also compared to experiments that were performed on sounding rocket TEXUS-42 [27, 28].

The influence of the shape of the channel's cross-section on the flow behaviour was investigated in further drop tower experiments. Critical flow rates were determined for a groove-shaped channel (i.e. parallel plates with one free surface walled off) and compared with the one-dimensional model by Haake et al. [30]. Klatte et al. [34] used the numerical tool Surface Evolver [7] to predict three-dimensional free surface profiles and flow rate limits of steady flow in a triangular wedge-shaped channel. The numerical predictions were in good agreement with those of the one-dimensional model adapted for a wedge-shaped duct and with the results of drop tower experiments [33]. Further drop tower experiments were performed by Wei et al. [62] using a similar triangular wedge-shaped channel. Their results confirm the findings of Klatte including the occurrence of flow separation at the free surface in CFD simulations.

Entering the regime of multiphase flow, Salim et al. [50] performed two-phase flow experiments in an open capillary channel on sounding rocket flight TEXUS-45. Bubbly flow was generated by means of a set of thin needles located upstream of the test channel. Based on their observations of the free surface profile in single-phase flow and two-phase bubbly flow, the authors concluded that wall shear stress increases when bubbles are injected into the flow stream.

So far, experiment data on open capillary channel flow has been thinly distributed across various channel geometries and flow regimes. Due to the restrictions of the experiment setups, large parametric studies were not possible outside of numerical simulations. Recently however, the CCF experiment on the International Space Station has generated a wealth of experiment data for three channel geome-

tries and various flow regimes. A selection of the experiment results are presented
and discussed in this thesis. Details of the experiment setup and the first results of
the flow rate limitation experiments are found in Canfield et al. [13] including quali-
tative and quantitative observations of the bubble formation process in supercritical
choked flow. A comprehensive review of the one-dimensional model for the open
capillary channel with a rectangular cross-section is given by Conrath et al. [19],
and its numerical results are compared with those gained from the ISS experiments
with special attention paid to the differences between fully developed flow and plug
flow at the inlet of the test channel. Bronowicki et al. [9] investigate the model for
open capillary channel flow through parallel plates. After validating the model by
comparing numerical results with experiment data the authors continue to study
the model's predictions of the shape of the free surface within the steady, subcritical
flow regime. Special attention is given to the transition between two flow regimes, in
which pressure loss within the channel is predominantly attributed to either convective or viscous effects, respectively. The one-dimensional model is expanded by Grah
et al. [29] to include transient effects and validated with experiments performed on
the ISS. Additional experiments on passive, geometry-based phase separation were
performed using the same setup and results are published by Weislogel et al. [64].

## 1.2.2   Multiphase Flow in Microgravity

One goal of this thesis is to determine the influence of a mono-disperse, gaseous second phase on the flow rate limitation that is evident in the single-phase experiments
involving liquid flowing through an open capillary channel. The behavior of two-
phase flows in closed pipes in microgravity has been the subject of various studies,
primarily to determine flow pattern transitions and also to examine the influence
of the second phase on the pressure drop within the channel. In both cases, the
microgravity environment is of great importance. The behaviour of two-phase gas-
liquid flows through pipes in normal gravity is influenced largely, if not dominated,
by body forces acting on the dispersed phase. In reduced or compensated gravity
environments, these forces are negligible and thus the variation of flow patterns and

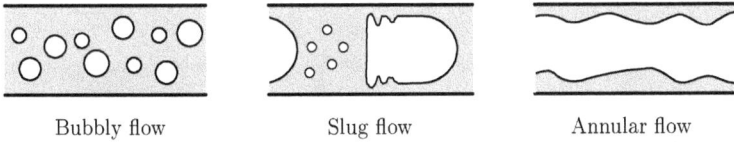

|              |            |              |
| Bubbly flow  | Slug flow  | Annular flow |

Figure 1.2: Generic sketches depicting typical two-phase flow patterns through a closed, circular channel under microgravity conditions and with flow from left to right. The disperse gaseous phase is coloured white, the continuous liquid phase is coloured grey.

the typical conditions at which the respective patterns occur differ to two-phase flow in normal gravity.

**Two-Phase Flow Patterns**

In two-phase flow in closed channels, the flow pattern has a strong influence on the pressure drop that occurs within the flow. Consequently, it is important to determine or predict the flow pattern that will occur in the experiment. Dukler et al. [23] were the first to conduct a large number of experiments in drop tower experiments and parabolic flights to determine a flow pattern map for air-water flow closed duct flow in a microgravity environment. Three distinct patterns were identified with transition regions between them: bubbly, slug, and annular flow. Various papers are dedicated to determining appropriate dimensionless numbers that can be used to adequately differentiate the flow patterns of two-phase flow in microgravity. For example, further investigations on flow patterns of two-phase flow in microgravity were conducted by Colin and Fabre [17] (using air and water), Bousman et al. [5], and Zhao and Rezkallah [71] to name but a few. Experiment data available at the time is summarized by Jayawardena et al. [32], and a dimensionless flow pattern map is proposed for two-phase flow in microgravity based solely on the respective Reynolds numbers of the phases (based on the respective superficial velocities and viscosities) and the Suratman number ($\mathrm{Su} = \mathrm{Re}_L^2/\mathrm{We}_L = \sigma d'_h \rho_L/\mu_L^2$). In response, Vasavada et al. [61] recap various models to predict the transition from bubbly to slug flow in microgravity and conclude that the Suratman number shows promise in this regard, but stress that the void fraction $\phi$ should also be taken into account. They base their conclusion on observations from Colin et al. [18], who argue that the flow pattern transition depends on the pipe size and the physical properties of the fluids but should be independent of the superficial velocities. Colin et al. [18] propose that a critical void fraction can be used to predict the transition between bubbly flow and slug flow where $\phi = 30\%$ for Su $< 1.7 \times 10^6$ and $\phi = 45\%$ for Su $> 1.7 \times 10^6$. It should be noted that the experiment study of Vasavada et al. [61] was performed at low Reynolds numbers, therefore inertia was not a dominant force in the two-phase flow. In the majority of the experimental studies on phase transition, high Reynolds numbers were prevalent and therefore turbulent, inertia-dominated flows were examined. Woelk et al. [67] carried out a number of drop tower experiments to compare two-phase flow patterns and their transitions in normal and microgravity environments.

Most authors agree that further work is required to determine an adequate means of predicting the flow pattern transitions over a wide range of flow regimes. While flow patterns and their transitions are not of primary interest in this thesis, the flow pattern does have an effect on the choice of an adequate mode and on the pressure drop characteristics and it is therefore important to describe the flow regime appropriately. The two-phase experiments discussed in this thesis are situated within the bubbly flow regime.

**Viscous Pressure Loss**

Another focus of two-phase flow experiments in microgravity, and of some importance for this thesis, is the comparison of the single-phase and the two-phase friction factors. While no single model currently exists to accurately predict pressure loss under microgravity conditions, various studies have been performed to either modify existing normal gravity models, propose new models, or create approximate correlations based on empirical data. The predictions vary in their accuracy depending on

various experiment conditions including, but not limited to, the diameter and shape
of the duct, the physical properties of the flowing phases, whether flow is laminar
or turbulent, the void fraction, and the flow pattern.

Colin and Fabre [17] found that viscous losses are underestimated by the single-
phase friction factor for turbulent flows at Reynolds numbers below 20,000 and
that the presence of large bubbles modifies the velocity profile in the liquid phase
which in turn appears to lead to an increase in wall shear stress. Furthermore, they
emphasize the necessity for additional experiments in the low Reynolds number
regime to further examine the wall shear stress in laminar bubbly flow.

Various models for predicting the viscous pressure loss are used throughout literature. Chen et al. [14] review various models for calculating the pressure drop
in two-phase flow under microgravity conditions and find that the Beattie-Whalley
model can be used for mono-disperse bubble flows. Beattie and Whalley [3] propose
that the two-phase flow can be modelled as a homogeneous mixture with an average
density and an average viscosity based on the void fraction within the flow. The
friction pressure gradient is then calculated using the viscous pressure loss equation
for single-phase flow albeit with the calculated mixture properties.

Zhao et al. [70] also perform experiments and compare other published results to
a modified homogeneous flow model for bubbly flow in microgravity. The majority
of their experiments were performed in turbulent flow. Their model accounts for
the observation that bubbles travel close to the axis in pipe flow which means that
wall friction is mainly based on the viscosity of the liquid. In addition, the authors
propose that a no-slip velocity condition is present in bubbly flows in microgravity
and that the two-phase friction factor should depend on the averaged mixture veloc-
ity $v'_m$ and the two-phase mixture Reynolds number $\mathrm{Re}_{2P} = \rho_L v'_m d'_h / \mu_L$. They find
a two-phase Fanning friction factor in laminar circular pipe flow of 35 $\mathrm{Re}_{2P}^{-1}$, which
is more than two times higher than the single-phase friction factor, and identify an
apparent transition to a higher friction factor in the range of $3000 < \mathrm{Re}_{2P} < 4000$
similar to the laminar-turbulent transition in single-phase flow. Their findings are

confirmed by Salim et al. [49, 50] who performed single-phase and bubbly flow experiments in an open capillary channel on a sounding rocket flight. Comparing the pressure drop of the two regimes, they conclude that the wall shear stress is significantly enhanced in the case of bubbly flow but also appeared to depend on the void fraction of the two-phase flow.

Choi et al. [15] also performed a study comparing bubbly flow at normal gravity and microgravity and came to the conclusion that the reduction of gravity does not influence the frictional pressure drop significantly in turbulent flow and can be effectively predicted by the well-known Blasius relation. In laminar flow, however, the maximum ratio of viscous pressure drop under Earth-gravity to that under microgravity is found to be 1:1.3.

In summary, a review of literature on this particular subject reveals that viscous pressure loss in bubbly, two-phase, laminar duct flow under microgravity conditions may be expected to be larger than under normal gravity conditions. A precise and generally accepted model that accurately predicts viscous pressure loss for all flow patterns in laminar two-phase flow remains to be established.

# Chapter 2

# Theory

## 2.1 Capillarity and Microgravity

In liquids, neighbouring molecules exert a van der Waals attraction on each other, where molecules of the same species have a higher attractive potential than molecules of different species. This attractive force is generally higher than the shorter ranged repulsive force between the same molecules and thus leads to cohesion within a liquid bulk and to adsorption or adhesion at liquid-solid boundaries. While the molecules in the bulk of a liquid will statistically have the same number of neighbouring molecules in each direction and therefore be exposed to a net zero-directional and stabilizing force, molecules at a gas-liquid interface experience an attraction potential difference normal to the interface due to the weaker attraction of the sparser distribution of molecules and possibly differing species in the gas phase. With respect to the liquid bulk, molecules along the liquid-gas interface will therefore be attracted inwards towards the liquid bulk and work must be done to establish the interface. Therefore, liquid molecules at an interface have a higher potential energy than liquid molecules in the liquid bulk [11]. As Sophocleous [55] states, "this [potential energy difference] is characterized by the surface tension, which is the interface potential energy divided by the interface area $(J/m^2)$, which manifests itself as a force per unit length $(N/m)$."

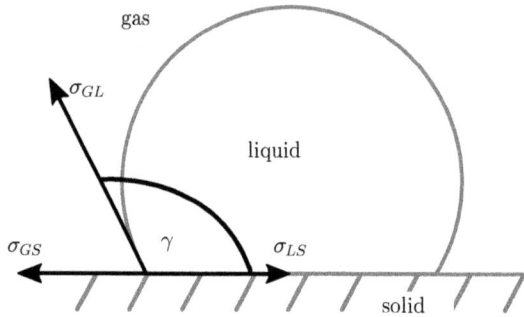

Figure 2.1: Contact line of three phases (here gas, liquid, solid). The macroscopic contact angle of the gas-liquid interface depends on the interfacial tension equilibrium between the three phases.

When a fluid-fluid interface (liquid-liquid or gas-liquid) comes into contact with a solid boundary, three equilibrium regimes are observed and are distinguished by the wettability of the liquid-solid material combination. Either (i) the liquid is found to be completely wetting on the solid boundary, (ii) the liquid is found to be partially wetting on the solid boundary, or (iii) the liquid is found to be partially non-wetting on the solid boundary. The three cases coincide with apparent macroscopic contact angles of $\gamma = 0°$, $90° \geq \gamma > 0°$, and and $180° > \gamma > 90°$ respectively[1], where $\gamma$ is the macroscopic contact angle. Thus, wettability refers to the liquid's tendency to spread over a solid and the contact angle is a manifestation of the relative forces of cohesion within the liquid and adhesion of the liquid to the solid boundary at a solid-fluid-fluid contact point. The macroscopic contact angle of two fluids at a solid boundary can be described as the sum of the individual tangential component of the interface tensions as outlined in figure 2.1. With $\sigma_{GL}$, $\sigma_{LS}$, and $\sigma_{GS}$ as the respective surface tensions at the gas-liquid, liquid-solid, and gas-solid interfaces, the equation describing the balance equation for the tangential stress components yields:

$$\sigma_{GS} = \sigma_{LS} + \sigma_{GL} \cos \gamma \quad .$$                                      (2.1)

---

[1]For $\gamma = 180°$ no wetting occurs, because the liquid does not in fact come into contact with the solid boundary.

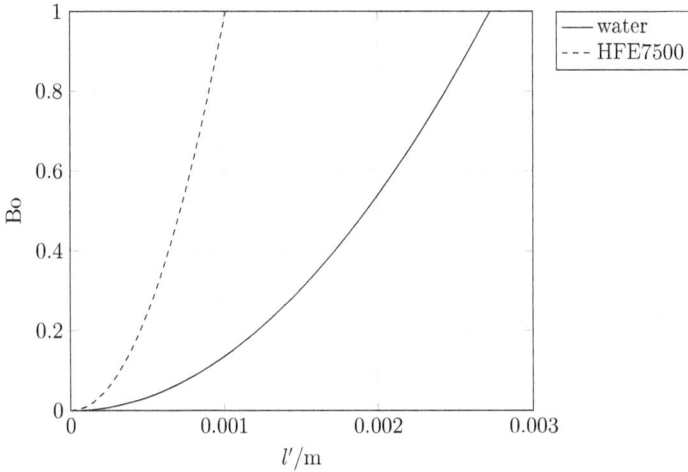

Figure 2.2: Length dependency of the ratio of hydrostatic pressure and capillary pressure, which is quantified by the Bond number.

Furthermore, when the liquid-gas interface is curved, a pressure difference across the interface is observed that is attributed to the surface tension force. This pressure difference, also referred to as capillary pressure, depends on the magnitude of the surface tension and on the radius of curvature of the interface and can be described with the Young-Laplace equation (compare equation (1.1)). The sign of the respective radii must be considered such that the additional pressure is pointed towards the centre of curvature.

Capillary pressure is generally overshadowed by hydrostatic pressure as displayed in figure 2.2. The ratio of gravitational body force to surface tension force is described by the dimensionless Bond number, Bo:

$$\text{Bo} = \frac{\Delta \rho g_i' L'^2}{\sigma_{GL}} \quad , \tag{2.2}$$

where $\Delta\rho$ is the difference between the densities of the two fluids[2], $g_i'$ is the acceleration (generally due to gravity) and $L'$ is a characteristic length scale. At very low Bo, capillary effects and surface tension dominate the behaviour of the gas-liquid interface. Generally, this physical regime is limited to short length scales. As can be seen in figure 2.2, hydrostatic and surface tension forces in water are in equilibrium (Bo = 1) when the characteristic length scale is in the order of a few millimetres. Therefore capillary effects can generally be neglected when dealing with fluid physics on Earth. While the length scale of microfluidic systems is often in the range of micrometres to millimetres, and thus small enough to be surface tension dominated, their flow regimes are limited to the viscous regime. In a reduced gravity environment such as on board a satellite or a space craft, however, the length scale of a capillary dominated fluidic device may be orders of magnitude higher than on Earth. This means that such flows are not limited to the viscous regime and inertia must also be considered when examining or modelling the behaviour of the flow and the gas-liquid interface.

Various experiment platforms exist with which microgravity may be achieved to investigate surface tension related phenomena. Meseguer et al. [40] list various methods to achieve microgravity and their respective levels of residual acceleration and their microgravity durations (compare table 2.1).

Table 2.1: Overview of microgravity duration and levels of residual acceleration for various microgravity experiment platforms.

| Experiment platform | Duration | residual acceleration |
|---|---|---|
| Drop towers | < 10 s | $10^{-3}g_i'$ to $10^{-6}g_i'$ |
| Aircraft | < 30 s | $10^{-2}g_i'$ to $10^{-3}g_i'$ |
| Sounding rockets | < 360 s | $10^{-4}g_i'$ to $10^{-6}g_i'$ |
| Satellites | months | $10^{-3}g_i'$ to $10^{-6}g_i'$ |
| International Space Station | years | $10^{-3}g_i'$ to $10^{-6}g_i'$ |

---

[2]When dealing with a gas-liquid interface, the density of the gas is often multiple orders of magnitude lower than that of the liquid and can therefore be neglected.

## 2.2    1D Model for Open Capillary Channel Flow

Consider forced flow through an open wedge-shaped channel with an isosceles triangular cross-section (compare figure 3.3). The flow is laminar, Newtonian, and isothermal. The channel geometry is constant and given by its length $l'$, its height $b$, and the angle $\alpha$, which is half the angle that is enclosed by the sides of the triangle with equal lengths. Edge $a$ is located opposite $\alpha$. The origin of the cartesian coordinate system is placed in the vertex of the triangular cross-section at the inlet of the test channel, which is defined as the cross-section that coincides with the upstream pinning edge. Within the region of interest $(0 < x' < l')$, edge $a$ is open to the gaseous atmosphere while the remaining edges are bounded by walls. For $x' \leq 0$ and for $x' \geq l'$, the channel is bounded on all three sides by walls. Pinning edges are located at $x' = 0$ and $x' = l'$, between which the liquid forms a deformable interface along edge $a$. It is well known that capillarity dominates the shape of the liquid-gas interface in a reduced gravity environment $(g_i' \to 0)$. The shape of the interface depends on the capillary pressure difference, which is described by the Young-Laplace equation:

$$\frac{p_l' - p_a'}{\sigma} = -2H' = -\left(\frac{1}{R_1'} + \frac{1}{R_2'}\right) \quad , \tag{2.3}$$

where $p_a'$ is the pressure in the ambient gas and considered constant, $p'$ is the pressure in the liquid, $\sigma$ is the surface tension of the liquid, and $H'$ is the mean curvature of the interface with the two principle radii $R_1'$ and $R_2'$.

The governing equations for the flow in the capillary channel are the conservation equations for mass and momentum. These equations are formulated in the following sections.

### 2.2.1    Conservation of Momentum

Flow in the open channel is modelled using the Navier-Stokes momentum equation [66] for an incompressible fluid with indices $i = \{1, 2, 3\}$ and $j = \{1, 2, 3\}$ indicating

the tensor dimension and flow direction with $\{1, 2, 3\}$ referring to the directions
$\{x', y', z'\}$ in the cartesian coordinate system , respectively:

$$\rho \left( \frac{\partial v'_i}{\partial t'} + v'_j \frac{\partial v'_i}{\partial x'_j} \right) = -\frac{\partial p'}{\partial x'_i} - \frac{\partial \tau'_{ji}}{\partial x'_j} + \rho g'_i \quad , \tag{2.4}$$

where $\rho$ is the density of the liquid, $v'_i$ is the velocity vector, $t'$ is time, the vector
$g_i$ describes body forces such as gravity, and $\tau'_{ji}$ is the stress tensor. The above
equation is further simplified in accordance with the following assumptions [58]:

a) Body forces are negligible due to the microgravity environment:

$$\rho g'_i = 0 \quad . \tag{2.5}$$

b) The flows considered in this work are all steady and transient effects are neglected:

$$\rho \frac{\partial v'_i}{\partial t'} = 0 \quad . \tag{2.6}$$

c) Flow is considered to be one-dimensional in character and streamline theory
is applicable.

d) The magnitudes of the velocity components that are normal to the direction
of flow are negligible (fully developed flow). Also, the average flow velocity is
defined as:

$$v'(x') = \frac{1}{A'(x')} \int_0^{A'(x')} v'_1(x', y', z') dA' \quad . \tag{2.7}$$

e) Cross flow is considered to be negligible.

f) The gas phase is passive in nature and no tangential viscous stress occurs at
the interface.

g) The model is limited to subcritical flows, i.e. $Q' \leq Q'_{crit}$.

h) The liquid is fully wetting with a static contact angle of $\gamma = 0°$.

i) The curvature of the interface has a symmetry plane in the $y'z'$-plane..

Based on the above assumptions, equation (2.4) simplifies to

$$-\frac{\mathrm{d}p'}{\mathrm{d}x'} = \rho v' \frac{\mathrm{d}v'}{\mathrm{d}x'} + \frac{\mathrm{d}w_f'}{\mathrm{d}x'} \quad , \tag{2.8}$$

with the viscous term

$$\frac{\mathrm{d}w_f'}{\mathrm{d}x'} = \frac{\partial \tau_{ji}'}{\partial x_j'} = \left(\frac{\partial \tau_{12}'}{\partial x_2'} + \frac{\partial \tau_{13}'}{\partial x_3'}\right) = -\mu \left(\frac{\partial^2 v_1'}{\partial x_2'^2} + \frac{\partial^2 v_1'}{\partial x_3'^2}\right) \quad . \tag{2.9}$$

Furthermore, according to equation (2.3), the pressure in the liquid can be described in terms of curvature:

$$\mathrm{d}p' = p' - p_a = -2\sigma \mathrm{d}H' = -\sigma \mathrm{d}h' \quad , \tag{2.10}$$

where $p_a$ is constant and $h' = 2H'$. Substituting equation (2.10) into (2.8) yields the governing momentum balance equation for the presented flow problem:

$$\sigma \frac{\mathrm{d}h'}{\mathrm{d}x'} = \rho v' \frac{\mathrm{d}v'}{\mathrm{d}x'} + \frac{\mathrm{d}w_f'}{\mathrm{d}x'} \quad . \tag{2.11}$$

The above equation shows that the capillary pressure gradient of the interface within the open capillary channel balances the pressure gradient caused by the sum of the convective and viscous terms of the momentum balance equation. This means that higher pressure losses in the channel are balanced by a greater curvature of the interface. In effect, the faster the liquid flows and/or the greater the viscous losses are, the more the interface must bend into the channel to balance the pressure difference between the flowing liquid and the ambient gas. At some point, however, the pressure loss in the flow induces a maximum curvature of the interface that cannot be exceeded due to the geometrical constraints of the channel. A further increase of the pressure loss in the flow would lead to an unbalanced equation (2.11) at which point the channel is considered to be choked.

## 2.2.2 Conservation of Mass

The general differential equation describing the conservation of mass (also called the continuity equation) is defined as [66, 68]:

$$\frac{\partial \rho}{\partial t'} + \frac{\partial (\rho v_i')}{\partial x_i'} = 0 \quad . \tag{2.12}$$

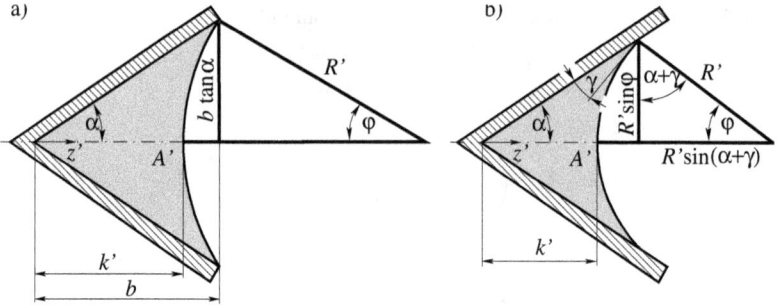

Figure 2.3: Contour $k'$, radius $R'$, and cross-sectional area $A'$ within the test channel and in the $y'z'$-plane for pinned (a) and unpinned (b) liquid interfaces. The liquid is coloured grey.

Based on the assumptions stated above and with the flow rate $Q' = v'A'$, the conservation of mass is described by the continuity equation for an incompressible, steady, one-dimensional flow with mean velocity $v'$ through cross-sectional area $A'$ [19, 33, 46], which yields:

$$0 = A'\frac{\mathrm{d}v'}{\mathrm{d}x'} + v'\frac{\mathrm{d}A'}{\mathrm{d}x'} \quad . \tag{2.13}$$

### 2.2.3   Area of cross-section

Within the test channel, the cross-section of the flow path in the $y'z'$-plane varies along $x'$ due to the pressure loss in the flow which is balanced by the curvature of the interface. The area of the cross-section is required to determine the average velocity along $x'$, which also varies in accordance with the conservation of mass as defined in equation (2.13). In order to determine the cross-sectional area one must distinguish two cases. Firstly, the capillary pressure is low and the liquid is pinned to the top of the walls. Secondly, the capillary pressure is high and the liquid interface is no longer pinned to the top edge of the walls. These two cases are sketched in figure 2.3 and the limit between them is defined here as a critical capillary pressure, which is attained when the interface describes a segment of a circle with the highest curvature that is possible given the geometrical constraints of the test channel. This critical capillary pressure can also be described in terms of critical contour height $k'_c$ when the assumptions listed in section 2.2.4 are taken into account. As displayed in figure 2.3, when $k' = k'_c$ the following relationship may be derived:

$$R' \sin \varphi = b \tan \alpha \quad . \tag{2.14}$$

With $\varphi = \pi/2 - (\alpha + \gamma)$ and $\sin \varphi = \cos(\alpha + \gamma)$ the radius $R'$ yields:

$$R' = \frac{b \tan \alpha}{\cos(\alpha + \gamma)} \quad , \tag{2.15}$$

and with $b - k'_c = R' - R' \sin(\alpha + \gamma)$, the critical contour height $k'_c$ between pinned and unpinned interface at the side wall can be described as follows:

$$k'_c = b \left( 1 - \frac{1 - \sin(\alpha + \gamma)}{\cos(\alpha + \gamma)} \tan \alpha \right) \quad . \tag{2.16}$$

Having determined the critical contour height that separates the pinned case from the non-pinned case, the cross-sectional area can also be described for either case as a function of $k'$ and the test channel geometry. This is accomplished by subtracting the circular segment above the free surface from the base area of the test channel's cross-section. For the pinned case the radius of the circular segment is defined as:

$$R' = \frac{(b \tan \alpha)^2 + (b - k')^2}{2(b - k')} \tag{2.17}$$

The cross-sectional area is then found by subtracting the circular segment above the free surface from the base area of the test channel which yields:

$$A' = (R' + k')b \tan \alpha - R'^2 \arcsin \frac{b \tan \alpha}{R'} \quad . \tag{2.18}$$

For the non-pinned case, the radius of the circular segment described by the interface in the $y'z'$-plane is defined as:

$$R' = k' \frac{\sin \alpha}{\cos \gamma - \sin \alpha} = k' \Psi \quad , \tag{2.19}$$

where $\Psi$ is a geometrical factor based on the contact angle $\gamma$ and the wedge's opening angle $\alpha$:

$$\Psi = \frac{\sin \alpha}{\cos \gamma - \sin \alpha} \quad . \tag{2.20}$$

The cross-sectional area for the unpinned case then yields:

$$A' = R'^2(\sin\alpha - \varphi) - R'k'\sin\alpha \quad . \tag{2.21}$$

In summary, the cross-section area of the streamtube may be described by the following function:

$$F'_A(k') = \begin{cases} (R' + k')b\tan\alpha - R'^2\arcsin\dfrac{b\tan\alpha}{R'} & ; \quad k' \geq k'_c \, , \\[2mm] (k'\Psi)^2[(1 + \dfrac{1}{\Psi})\sin\varphi - \varphi] & ; \quad k' < k'_c \, , \end{cases} \tag{2.22}$$

where $k' \geq k'_c$ is valid when the liquid is pinned to the top of the side walls of the channel and $k' < k'_c$ is valid when the liquid is not pinned.

### 2.2.4   Mean Curvature of the Interface

The curvature of a surface $z' = z'(x', y')$ is described by [10, 42]

$$H' = \frac{\dfrac{\partial^2 z'}{\partial x'^2}\left[1 + \left(\dfrac{\partial z'}{\partial y'}\right)^2\right] - 2\dfrac{\partial z'}{\partial x'}\dfrac{\partial z'}{\partial y'}\dfrac{\partial^2 z'}{\partial x'\partial y'} + \dfrac{\partial^2 z'}{\partial y'^2}\left[1 + \left(\dfrac{\partial z'}{\partial x'}\right)^2\right]}{2\left[1 + \left(\dfrac{\partial z'}{\partial x'}\right)^2 + \left(\dfrac{\partial z'}{\partial y'}\right)^2\right]^{3/2}} \quad . \tag{2.23}$$

Furthermore, let us assume that the minimal value of $k'$ for each cross-section normal to $x'$ is located at $y' = 0$ and that the curvature of the area in each cross-section normal to $x'$ is constant with a symmetry plane located in the $x'z'$-plane. From these assumptions we find that we can define the contour of the surface $z'(x', y')$ in the $x'z'$-plane as $k'(x') = z'(x', y' = 0)$. We find also that due to symmetry $\partial z'/\partial y' = 0$ in the $x'y'$-plane. This simplifies equation (2.23) to

$$H' = \underbrace{\frac{\dfrac{\partial^2 k'}{\partial y'^2}}{2\left[1 + \left(\dfrac{\partial k'}{\partial x'}\right)^2\right]^{1/2}}}_{0.5\,R'^{-1}_1} + \underbrace{\frac{\dfrac{\partial^2 k'}{\partial x'^2}}{2\left[1 + \left(\dfrac{\partial k'}{\partial x'}\right)^2\right]^{3/2}}}_{0.5\,R'^{-1}_2} \quad , \tag{2.24}$$

where the first term represents the first radius of curvature, $R'_1$, and the second term represents the second radius of curvature, $R'_2$. Finally, $H'$ is substituted in accordance with equation (2.10), which yields

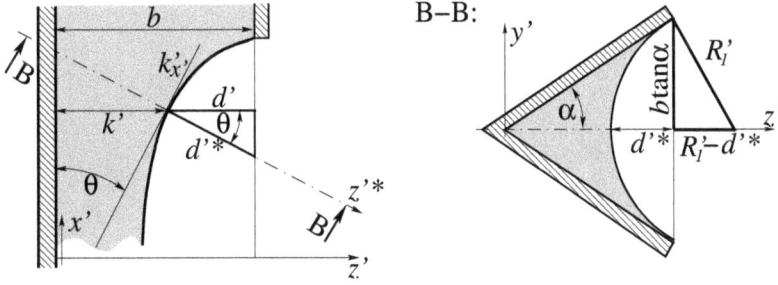

Figure 2.4: Inlet section of the test channel in the $x'z'$-plane (left) and cross-section of the test channel in the plane that is perpendicular to the interface and inclined to the $y'z'$-plane at angle $\theta$ (right). The liquid is coloured grey.

$$h' = \frac{1}{R_1'} + \frac{\dfrac{\partial^2 k'}{\partial x'^2}}{\left[1 + \left(\dfrac{\partial k'}{\partial x'}\right)^2\right]^{3/2}} \quad . \tag{2.25}$$

The radii of curvature are positioned perpendicular to each other and thus $R_1'$ is positioned in the $y'z'$-plane only when $\partial k'/\partial x' = 0$. Otherwise, the first principle radius of curvature, $R_1'$ is positioned at an inclined angle $\theta$ to the $y'z'$-plane (compare figure 2.4) and is therefore not equal to the radius of the cross-sectional area, $R'$. With respect to the inclination of $\theta$, we can define a length $d'^*$ as follows:

$$d'^* = d'\left[1 + \left(\frac{\partial k'}{\partial x'}\right)^2\right]^{1/2} = (b - k')\left[1 + \left(\frac{\partial k'}{\partial x'}\right)^2\right]^{1/2} \quad , \tag{2.26}$$

with $d'^* = b - k'$ for $\theta = 0$. Applying Pythagoras' theorem to the tilted plane one finds:

$$R_1'^2 = (b\tan\alpha)^2 + (R_1' - d'^*)^2 \quad . \tag{2.27}$$

In combination with equation (2.26), the first principal radius of curvature yields (for the pinned case with $k' \geq k'_{c1}$):

$$R'_1 = \frac{(b\tan\alpha)^2 + (b - k')^2 \left[1 + \left(\frac{\partial k'}{\partial x'}\right)^2\right]}{2(b - k')\left[1 + \left(\frac{\partial k'}{\partial x'}\right)^2\right]^{1/2}} \quad . \tag{2.28}$$

For the unpinned case with $k' < k'_{c1}$, the first principal radius of curvature yields:

$$R'_1 = k'\frac{\sin\alpha}{\sin\gamma - \sin\alpha} \quad , \tag{2.29}$$

with the the transition from pinned to unpinned at $k'_{c1}$ that is defined according to equation (2.16):

$$k'_{c1} = b\left(1 - \frac{1 - \sin(\alpha + \gamma)}{\cos(\alpha + \gamma)}\tan\alpha\left[1 + \left(\frac{\partial k'}{\partial x'}\right)^2\right]^{1/2}\right) \quad . \tag{2.30}$$

## 2.2.5   Irreversible Pressure Loss

Flow through the open test channel is still subject to viscous pressure loss due to friction along the side walls. Ayyaswamy et al. [2] determined numerical solutions for the friction coefficient of fully developed steady laminar flow through liquid filled triangular grooves with a shear free meniscus. Their results are presented in tabulated form for contact angles $0.1° \leq \gamma \leq 90° - \alpha$ and for channel half angles $5° \leq \alpha \leq 60°$. With the general *ansatz* [66] for viscous pressure loss

$$\Delta p' = \frac{K_f}{\text{Re}}\frac{\Delta x'}{d'_h}\frac{\rho v'^2}{2} \quad , \tag{2.31}$$

and with Ayyaswamy's definition for the friction factor $K_f$

$$K_f = \frac{2D^2_{Ayy}}{\bar{v}_{Ayy}} = \frac{2d''^2_h}{\bar{v}_{Ayy}r'^2_W} \quad , \tag{2.32}$$

where $\bar{v}_{Ayy}$ is a dimensionless velocity. $D_{Ayy}$ is a dimensionless hydraulic diameter and $d'_h$ is a dimensional hydraulic diameter with $d'_h = 4A'/P'$, where $P'$ is the wetted perimeter. Re is the Reynolds number $\text{Re} = (d'_h v')/\nu$. Substitution into equation (2.11) yields:

$$\frac{dw'_f}{dx'} = \frac{\mu}{\bar{v}_{Ayy}r'^2_W}v' \quad . \tag{2.33}$$

The dimensionless velocity $\bar{v}_{Ayy}$ is a function of $\alpha$ and the effective contact angle $\gamma_{eff}$ between the free surface and the side wall of the channel, and $r_W'^2$ is the length of the wetted side wall. The effective contact angle is defined as:

$$\gamma_{eff} = \begin{cases} \arccos\left(\dfrac{b\tan\alpha}{R'}\right) - \alpha & \text{if } k' \geq k_c' \quad, \\ \gamma & \text{if } k' < k_c' \quad, \end{cases} \tag{2.34}$$

and the length of the wetted side wall is defined as:

$$r_W' = \begin{cases} \dfrac{b}{\cos\alpha} & \text{if } k' \geq k_c' \quad, \\ R'\dfrac{\cos(\alpha+\gamma)}{\sin\alpha} & \text{if } k' < k_c' \quad. \end{cases} \tag{2.35}$$

Tabulated results of the cited study are presented in appendix C and the friction coefficients for fully developed flow are interpolated using a cubic spline for given $\gamma_{eff}$ and $\alpha$.

## 2.2.6  Boundary Conditions

In summary, the presented flow problem is described using the three coupled partial differential equations (2.11), (2.25), (2.13), and either algebraic equation (2.18) or equation (2.21). Four boundary conditions must be chosen to produce a solution. These boundary conditions are:

1) The average velocity at the outlet of the test channel is defined as the forced flow rate $Q'(t')$ divided by the base cross-sectional area of the test channel $A_0'$:

$$v'(x' = l') = \frac{Q'(t')}{A_0'} \quad. \tag{2.36}$$

2) The capillary pressure at the inlet of the test channel is known (details of how the capillary pressure is determined a priori are found in section 2.2.7):

$$h'(x' = 0) = h_0' \quad. \tag{2.37}$$

3) The free surface is pinned to the base wall at the inlet of the test channel. Therefore the height of the contour at the inlet of the open channel is

$$k(x' = 0) = b \quad. \tag{2.38}$$

4) The free surface is pinned to the base wall at the outlet of the test channel. Therefore the height of the contour at the outlet of the open channel is

$$k'(x' = l') = b \quad . \tag{2.39}$$

## 2.2.7 Pressure Boundary Condition

The pressure boundary condition at the inlet of the test channel ($x = 0$) is required for the one-dimensional model. An analytical derivation of the pressure boundary condition at the inlet of the test channel is impossible due to the complex geometry of the FPC and the entrance section that are located upstream of the test channel. Instead, three-dimensional computational fluid dynamics (CFD) simulations were performed to determine the pressure loss from the flow preparation chamber (FPC, see figures 3.5 and 3.6) to the inlet of the test channel as a function of the Reynolds number. The results of the simulations were used to approximate the pressure boundary condition at the inlet of the test channel using a polynomial function of first order. The pressure boundary condition is approximated assuming the following conditions:

a) flow is steady and laminar;

b) streamline theory is applicable;

c) the velocity of the liquid within the FPC is close to zero or negligibly small in comparison to the mean velocity of the liquid in the test channel.

Based on the above assumptions, the pressure at the inlet of the test channel is determined by following the pressure evolution along a streamline from the FPC to $x = 0$:

$$p'_{FPC} + \frac{1}{2}\rho v'^2_{FPC} = p'_0 + \frac{1}{2}\rho v'^2_0 + p'_n \quad , \tag{2.40}$$

where $p'_n$ is the irreversible pressure loss along the streamline. With $A'_0/A'_{FPC} = 0.017$ and due to conservation of mass, $v_{FPC} \ll v'_0$. Therefore, $v'_{FPC}$ is considered to be negligible in this balance equation.

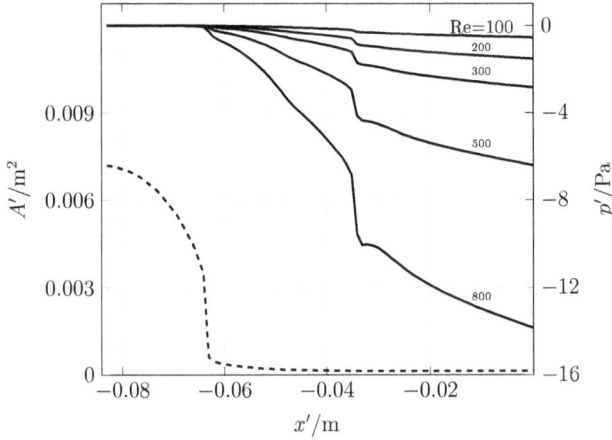

Figure 2.5: Pressure loss from the FPC to the inlet of the test channel. Solid lines display the pressure evolution through the entrance region according to CFD simulations at the respective Reynolds numbers. The dashed line represents the variation of the cross-sectional area in flow direction (compare figure 3.6).

The pressure in the FPC is defined by the ambient pressure and the capillary pressure in the compensation tube:

$$p'_{FPC} = p'_a - \frac{2\sigma}{R'_{CT}} \quad . \tag{2.41}$$

Rearranged and with the curvature at the inlet of the test channel $\sigma h'_0 = p'_a - p'_0$, equation (2.40) yields:

$$\sigma h'_0 = \frac{2\sigma}{R'_{CT}} + \frac{1}{2}\rho v_0'^2 + p'_n \quad . \tag{2.42}$$

Three-dimensional CFD simulations were performed at various Reynolds numbers ($50 < \text{Re}_c < 800$)[3] to determine solutions for $p'_n$ (compare section 4.2). The results are displayed in figure 2.5. The pressure loss increases with Reynolds number and depends on the cross-sectional area of the flow path. Of special note is the

---

[3]In this case, Re is based on the hydraulic diameter of the closed channel, which is defined as $d'_{hc} = 2b(1 + \sin^{-1}\alpha)^{-1}$.

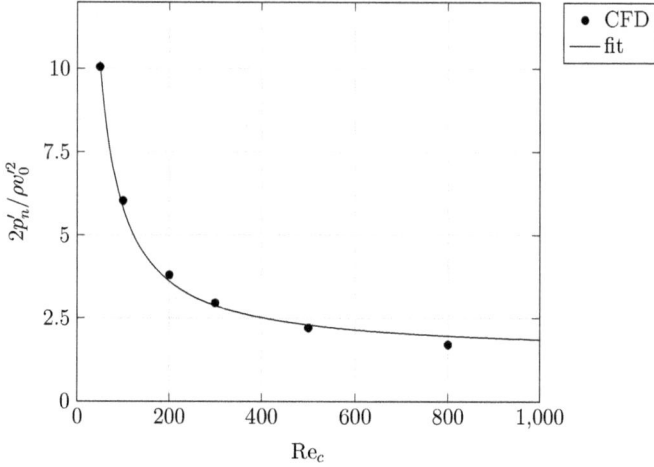

Figure 2.6: Scaled irreversible pressure loss over Reynolds number ($\text{Re}_c = v_0' d_{hc}' \nu^{-1}$). Results are derived from three-dimensional CFD simulations of flow through the FPC, nozzle and closed test channel.

significant pressure drop in all cases at $x' \approx -0.034$ m. This is where the bubble injector is located. The bubble injector is a cannula that enters the channel from the base of the triangle and is bent into the flow path to fit into the apex of the triangular cross-section. The needle has an outer diameter of 0.8 mm and restricts the flow path by $A_{BI}'/A_0' \approx 0.2$. The flow path constriction causes an additional pressure loss that is accounted for in the pressure loss CFD simulations and is therefore also implemented in the pressure boundary condition of the 1D model. The irreversible pressure loss is plotted in figure 2.6 as a function of the Reynolds number and can be adequately approximated using a first polynomial function:

$$\frac{2p_n'}{\rho v_0'^2} = C_0 + \frac{C_1}{\text{Re}_c} \quad . \tag{2.43}$$

Inserting equation (2.43) into (2.42) yields

$$\sigma h_0' = \frac{2\sigma}{R_{CT}'} + (1 + C_0)\frac{\rho}{2}v_0'^2 + C_1\frac{\rho\nu}{2d_{hc}'}v_0' \quad . \tag{2.44}$$

The numerical solver ccFlow uses the following equation to compute the pressure boundary condition:

$$p'_a - p'_0 = \sigma h'_0 = K_0 \frac{2\sigma}{a} + K_1 \frac{\rho\nu}{a}\frac{v'_0}{4} + K_2 \frac{\rho v_0^2}{2} \quad . \tag{2.45}$$

Combining equations (2.44) and (2.45) yields the following relationship:

$$K_0 = \frac{a}{R'_{CT}} \quad ,$$

$$K_1 = C_1 \frac{2a}{d'_{hc}} \quad ,$$

$$K_2 = C_0 + 1 \quad . \tag{2.46}$$

CFD simulations are evaluated to determine the pressure loss as a function of the Reynolds number. Using the fit coefficients $C_0 = 1.415$ and $C_1 = 439$ and the hydraulic diameter of the closed channel $d'_{hc} = 7.25 \times 10^{-3}$ m, the coefficients in equation (2.46) amount to $K_0 = 0.278$, $K_1 = 1008$, and $K_2 = 2.415$. These are used to approximate the pressure boundary condition at the inlet of the test channel.

## 2.3   Modelling Two-Phase Flow

Within the scope of this thesis, multiphase flow is considered for isothermal, laminar, two-component, mono-disperse bubbly, gas-liquid flow through an open capillary channel in a microgravity environment. In addition to investigating the flow problem experimentally, an attempt is made to incorporate the physics of two-phase flow into the 1D model described in section 2.2 and to produce numerical predictions for the two-phase critical flow rate. The gaseous phase is injected at the inlet of the test channel in the vicinity of the vertex and is distributed within the channel as a mono-disperse bubble stream. In this thesis, the homogeneous flow model is applied to modify the model for single-phase capillary channel flow presented in the previous sections. The homogeneous flow model is a convenient choice in this case, because the adaptation of to the existing programme code is less extensive. In contrast to earthbound multiphase flows, relatively few studies were performed on

two-phase pressure loss in laminar flows in microgravity. According to the few that
are available, however, the homogeneous model appears to bear satisfactory results
for predicting pressure loss [24, 50, 70], though it should be noted that most studies
were conducted at much higher Reynolds numbers than observed here. Indeed,
Colin [16] points out that additional studies are required to adequately determine
the behaviour of multiphase flows in microgravity at low Reynolds numbers.

The homogeneous flow model considers the flow on a macroscopic level and
treats the two-phase mixture as a single-phase pseudo-fluid. This means that the
governing equations of the individual phases are not considered separately. Instead,
a single set of equations is required, as is the case in single-phase flow, albeit with
averaged velocity, temperature, density, and viscosity based on the mass fraction
of the phases. In addition, the relative velocity between the phases is neglected.
This model is mainly applied to two-phase flows in which the volumetric ratio of the
two phases is approximately constant in space such as in emulsions, foams, vapour
in turbines, etc. As a first step, an appropriate method of averaging the time-
dependent mixture properties must be chosen. The void fraction, $\beta$, is defined here
as the time-averaged fraction of the channel's volume that is occupied by the gaseous
phase[4]. The volume of the channel varies depending on the contour of the interface
as shown in section 2.2. Therefore, the time-averaged volumetric flow rates of liquid
and gas are used to determine a total flow rate, which yields the void fraction within
the channel:

$$\beta = \frac{\left(\dfrac{\partial V'_G}{\partial t'}\right)}{\left(\dfrac{\partial V'_G}{\partial t'}\right) + \left(\dfrac{\partial V'_L}{\partial t'}\right)} = \frac{Q'_G}{Q'_G + Q'_L} \quad , \tag{2.47}$$

where the subscripts $G$ and $L$ pertain to the gaseous phase and the liquid phase
respectively. The volume fraction of the liquid is then simply $(1 - \beta)$. Now that
the volumetric ratio is defined, the physical properties of the two phases can be

---

[4]In one-dimensional bubbly flow with zero slip velocity between the phases, the phase fraction
may be assumed to be equal the volumetric fraction [8]. Due to the assumption of negligible slip
velocity, the phrases 'volume fraction' and 'void fraction' may be considered synonymous in this
work.

merged to determine the modelled mixture's density assuming that both phases are incompressible:

$$\rho_m = \beta \rho_G + (1 - \beta)\rho_L \quad . \tag{2.48}$$

In this thesis, the volumetric flux (volume flow per unit area) is defined as:

$$u'_{Gi} = \beta v'_{Gi} \quad , \tag{2.49}$$
$$u'_{Li} = (1 - \beta)v'_{Li} \quad , \tag{2.50}$$

with a total volumetric flux that is defined as:

$$u'_i = u'_{Gi} + u'_{Li} \quad . \tag{2.51}$$

The relative velocity between the phases is denoted as $v'_{LGi}$ and is defined as:

$$v'_{LGi} = v'_{Li} - v'_{Gi} \quad . \tag{2.52}$$

Volumetric flux and velocity are related to each other such that:

$$u'_i = \beta v'_{Gi} + (1 - \beta)v'_{Li} \quad . \tag{2.53}$$

## 2.3.1   Conservation of Mass

Using the above notation, and following Brennen [8], a combined continuity equation may be formulated for the two-phase flow which will be subsequently simplified using basic assumptions. The two-phase combined continuity equation yields:

$$\frac{\partial}{\partial t'}\left[\beta \rho_G + (1 - \beta)\rho_L\right] + \frac{\partial}{\partial x'_i}\left(\rho_G u'_{Gi} + \rho_L u'_{Li}\right) = 0 \quad . \tag{2.54}$$

With equations (2.48) and (2.53) and assuming $v'_{LGi} = 0$ and thus using an averaged stream-wise mixture velocity $v'_{mi} = v'_{Li} = v'_{Gi}$, the above equation simplifies to a form identical to the continuity equation for single-phase flow (compare equation (2.12)) albeit using the mixture density:

$$\frac{\partial \rho_m}{\partial t'} + \frac{\partial}{\partial x'_i}(\rho_m v'_{mi}) = 0 \quad . \tag{2.55}$$

Assuming one-dimensional flow and that $\beta$ is constant throughout the channel, the averaged mixture velocity may be defined as:

$$v'_m = \frac{Q'_G + Q'_L}{A'} \quad . \tag{2.56}$$

Further simplification of the balance equation is based on the assumptions stated in section 2.3 and the final equation for conservation of mass for one-dimensional, incompressible, steady state, laminar, isothermal two-phase flow yields:

$$0 = \rho_m \left( A' \frac{\mathrm{d}v'_m}{\mathrm{d}x'} + v'_m \frac{\mathrm{d}A'}{\mathrm{d}x'} \right) \quad , \tag{2.57}$$

which is identical to the continuity equation for single-phase flow in equation (2.13) except that averaged mixture properties and a mixture velocity are used here.

## 2.3.2   Conservation of Momentum

The two-phase momentum balance equation is defined based on the same assumptions as the two-phase continuity equation. Conservation of momentum in steady, one-dimensional, two-phase flow with $\beta$ = constant throughout the channel may thus be described by the differential equation [1, 8, 49]:

$$\frac{1}{A'} \frac{\partial}{\partial x'} \left( \frac{\rho_G}{\beta} \frac{Q'^2_G}{A'} + \frac{\rho_L}{1-\beta} \frac{Q'^2_L}{A'} \right) = -\frac{\partial p'}{\partial x'} - \frac{P'}{A'} \tau'_w \quad . \tag{2.58}$$

Considering the aforementioned assumption of constant $\beta$ along the channel and with equations (2.48) and (2.56), the momentum equation is almost identical to the single-phase momentum equation except for the use of averaged properties and a mixture velocity and a two-phase pressure loss term on the right hand side of the equation. In this thesis, two-phase viscous pressure loss is calculated identically to single-phase flow albeit using $\rho_m$ and $\mu_L$ based on the assumption that the bubbly flow does not significantly influence the flow field and that viscous forces arise solely from the interaction of liquid and solid because of the wetting properties of the test liquid. Further development of the two-phase model is also identical in fashion to the single-phase model outlined previously except for the determination of the pressure boundary condition at the inlet of the test channel, which is presented

below. Ultimately, two-phase flow is modelled here using a homogeneous flow model
with mixture properties and velocities based on volumetric phase-weighted averages.

### 2.3.3   Pressure Boundary Condition

Bubble injection occurs at the inlet of the test channel, which means that the flow
upstream of the inlet is always single-phase. This must be accounted for in the
determination of the pressure boundary condition. The total flow rate $Q'$ is constant
and set by the pump. The flow rate in the entrance region upstream of the inlet
must therefore yield $Q' - Q'_G = Q'_L$. Equation (2.45) must be modified accordingly
and yields:

$$p'_a - p'_0 = \sigma h'_0 = K_0 \frac{2\sigma}{a} + K_1 \frac{\rho_L \nu_L}{a} \frac{(Q'_L)}{4A'_0} + K_2 \frac{\rho_L (Q'_L)^2}{2A'^2_0} \quad . \tag{2.59}$$

An additional modification of the boundary condition may be made (besides
the reduction of the liquid flow rate in the entrance region) to account for the
additional influx of momentum that stems from the gas injection at the inlet of
the test channel. Assuming that all injected bubbles are of equal and constant
volume, the displacement of the liquid in the vicinity of the injection site due to
the initial bubble growth is modelled as an additional inertial force based on the
linearized displacement velocity $(\Delta R'_{bi}/\Delta t'_{dc})$ which is defined via the duration of
bubble injection, $\Delta t'_{dc}$, and the ultimate volume of the injected bubbles with the
bubble radius $R'_{bi}$:

$$\Delta R'_{bi} = \left( \frac{3V''_b i}{4\pi} \right)^{1/3} \quad . \tag{2.60}$$

The additional pressure difference $(\Delta p'_{am})$ at the inlet of the TC due to the added
momentum in the liquid phase during displacement from bubble growth is defined
here as:

$$\Delta p'_{am} = \frac{1}{2} \rho_L \left( \frac{\mathrm{d} R'_{bi}}{\mathrm{d} t'} \right)^2 f'_{bi} t'_{dc} \quad , \tag{2.61}$$

where $f'_{bi}$ is the bubble injection frequency and $t'_{dc}$ is the duty cycle of the gas inlet
valve that is chosen as a characteristic time scale for individual bubble injections.

Table 2.2: Dimensionless quantities and characteristic numbers used in the scaling procedure for the wedge.

| Lengths | Other | Constants | Dimensionless numbers |
|---|---|---|---|
| $x = \dfrac{x'}{l'}$ | $v = \dfrac{v'}{v_c}$ | $v_c = \sqrt{\dfrac{\sigma}{\rho b \tan \alpha}}$ | $\tilde{l} = \dfrac{\mathrm{Oh} l'}{2 d'_h}$ |
| $y = \dfrac{y'}{b \tan \alpha}$ | $t = \dfrac{t' v_c}{l'}$ | $p_c = \dfrac{\sigma}{b \tan \alpha}$ | $\mathrm{Oh} = \sqrt{\dfrac{\rho \nu^2}{\sigma d'_h}}$ |
| $z = \dfrac{y'}{b \tan \alpha}$ | $p = \dfrac{p'}{p_c}$ | $\Gamma = \dfrac{b \tan \alpha}{l'}$ | $\Lambda = \dfrac{1}{\tan \alpha}$ |
| $k = \dfrac{k'}{b \tan \alpha}$ | $A = \dfrac{A'}{A'_0}$ | $A'_0 = b^2 \tan \alpha$ | |
| $h = h' b \tan \alpha$ | $Q = \dfrac{Q'}{A'_0 v_c}$ | $d'_h = 2b \sin \alpha$ | |
| $R = \dfrac{R'}{b \tan \alpha}$ | | $\mathrm{Re} = \dfrac{v_c b \tan \alpha}{\nu}$ | |
| $R_1 = \dfrac{R'_1}{b \tan \alpha}$ | | | |
| $R_2 = \dfrac{R'_2}{b \tan \alpha}$ | | | |
| $r_W = \dfrac{r'_W \cos \alpha}{b}$ | | | |

Combining equations (2.59) and (2.61) then yields the final pressure boundary condition at the inlet of test channel for two-phase flow with gas injection:

$$\sigma h'_0 = K_0 \frac{2\sigma}{a} + K_1 \frac{\rho_L \nu_L}{a} \frac{(Q'_L)}{4 A'_0} + K_2 \frac{\rho_L (Q'_L)^2}{2 A'^2_0} + \rho_L \left( \frac{\mathrm{d} R'_{bi}}{\mathrm{d} t'} \right)^2 f'_{bi} t'_{dc} \quad . \qquad (2.62)$$

## 2.4   The Non-Dimensionalized Model

The scaling of the presented flow problem is adapted from previous work on the identical flow problem in rectangular capillary channels [27, 30, 46]. A similar scaling method for the triangular channel has been recently published by Klatte [33], who couples the characteristic length scale to the maximum capillary pressure for a pinned interface within the channel. By doing so, he elegantly accounts for the inclination of the side walls of the triangular channel, a geometrical attribute that the wedge does not share with rectangular channels. Nonetheless, the scaling method for the rectangular channels is adapted and applied in this work. This is mostly due to convenience since the the software (ccflow) used to solve the one-dimensional model is programmed using this particular scaling method. The scaled variables are summarized in table 2.2.

Lengths are scaled by $b \tan \alpha$, $b/\cos \alpha$, or $l'$ depending on their orientation. The characteristic capillary pressure is defined as $p_c = \sigma/(b \tan \alpha)$ and the characteristic velocity is defined as $v_c = \sqrt{p_c/\rho}$. The characteristic numbers in the model are the Ohnesorge number:

$$\mathrm{Oh} = \sqrt{\frac{\rho \nu^2}{\sigma d'_h}} \quad , \tag{2.63}$$

the dimensionless channel length $\tilde{l}$:

$$\tilde{l} = \frac{\mathrm{Oh}\, l'}{2 d'_h} \quad , \tag{2.64}$$

and the aspect ratio $\Lambda$:

$$\Lambda = \frac{1}{\tan \alpha} \quad . \tag{2.65}$$

In addition, the static contact angle $\gamma$ is assumed to be both constant and zero in this model.

Equations (2.11), (2.13), (2.22), and (2.25) describe the one-dimensional model and are presented here again in dimensional form:

$$0 = -\sigma \frac{\mathrm{d}h'}{\mathrm{d}x'} + \rho v' \frac{\mathrm{d}v'}{\mathrm{d}x'} + \frac{\mu}{\bar{v}_{Ayy} r'^2_W} v' \quad , \tag{2.66}$$

$$0 = A' \frac{\mathrm{d}v'}{\mathrm{d}x'} + v' \frac{\mathrm{d}A'}{\mathrm{d}x'} \quad , \tag{2.67}$$

$$h' = \frac{1}{R'_1} + \frac{\dfrac{\mathrm{d}^2 k'}{\mathrm{d}x'^2}}{\left[ 1 + \left( \dfrac{\mathrm{d}k'}{\mathrm{d}x'} \right)^2 \right]^{3/2}} \quad , \tag{2.68}$$

$$0 = A' - F'_A(k') \quad . \tag{2.69}$$

Applying the presented scaling to the above equations yields the scaled one-dimensional model for steady open capillary channel flow. The respective dimensionless equations for momentum, continuity, curvature, and cross-sectional area are:

$$0 = -\frac{dh}{dx} + v\frac{dv}{dx} + \frac{K_f\,\tilde{l}}{2}\,v \quad, \tag{2.70}$$

$$0 = v\frac{dA}{dx} + A\frac{dv}{dx} \quad, \tag{2.71}$$

$$0 = \Gamma^2\frac{d^2k}{dx^2} + \left(\frac{1}{R_1} - h\right)\left[1 + \Gamma^2\left(\frac{dk}{dx}\right)^2\right]^{3/2} \quad, \tag{2.72}$$

$$0 = A - F_A(k) \quad. \tag{2.73}$$

Similarly, the dimensionless critical height that defines the border between pinned and unpinned interface is:

$$k_c = \frac{1}{\tan\alpha} - \frac{1 - \sin(\alpha + \gamma)}{\cos(\alpha + \gamma)} \quad. \tag{2.74}$$

The dimensionless equation for the cross-sectional area of flow is determined according to table 2.2 and equations (2.18) and (2.21):

$$F_A(k) = \begin{cases} \left(k + R - R^2\arcsin\frac{1}{R}\right)\tan\alpha & ;\ k \geq k_c \quad, \\ (k\Psi)^2\left[\left(1 + \frac{1}{\Psi}\right)\sin\varphi - \varphi\right]\tan\alpha & ;\ k < k_c \quad. \end{cases} \tag{2.75}$$

with the dimensionless geometrical factor $\Psi$ defined in equation (2.20), the pinning limit $k_c$ already defined in equation (2.74), and the angle $\varphi = \pi/2 - (\alpha + \gamma)$. The non-dimensional radius of the cross-sectional area is determined by scaling equations (2.17) and (2.19):

$$R = \begin{cases} \dfrac{\tan^2\alpha + (1 - k\tan\alpha)^2}{2(1 - k\tan\alpha)\tan\alpha} & ;\ k \geq k_c \quad, \\ k\dfrac{\sin\alpha}{\cos\gamma - \sin\alpha} & ;\ k < k_c \quad. \end{cases} \tag{2.76}$$

The dimensionless first principal radius of curvature yields according to equations (2.28) and (2.29):

$$
R_1 = 
\begin{cases}
\dfrac{\tan^2\alpha + (1-k)^2\left[1 + \Gamma^2\left(\dfrac{\partial^2 k}{\partial x^2}\right)^2\right]}{2(1-k)\tan\alpha\left[1 + \Gamma^2\left(\dfrac{\partial^2 k}{\partial x^2}\right)^2\right]^{1/2}} & ; \ k \geq k_{c1} \ , \\[4ex]
\dfrac{k\cos\alpha}{\cos\gamma - \sin\alpha} & ; \ k < k_{c1} \ .
\end{cases}
\tag{2.77}
$$

The critical height at which the interface transitions from a pinned interface to an unpinned one is defined in equation (2.30) and yields after scaling:

$$
k_{c1} = \left(\frac{1}{\tan\alpha} - \frac{1-\sin(\alpha+\gamma)}{\cos(\alpha+\gamma)}\right)\left[1 + \Gamma^2\left(\frac{\partial^2 k}{\partial x^2}\right)^2\right]^{-1/2} \ . \tag{2.78}
$$

The friction factor in the momentum equation is modelled according to equation (2.33) and reads in non-dimensionalized form:

$$
K_f = \frac{8\sqrt{2}\cos^3\alpha\sin^2\alpha}{\bar{v}_{Ayy}r_W^2} \ . \tag{2.79}
$$

The dimensionless velocity $\bar{v}_{Ayy}$ is approximated using a cubic spline interpolation applied to results from Ayyaswamy et al. [2] (compare appendix C) as a function of the opening angle $\alpha$ and the effective angle $\gamma_{eff}$ which is defined as the actual angle between the channel wall and the liquid interface (for the unpinned case, $\gamma_{eff} = \gamma$):

$$
\gamma_{eff} = \arccos\frac{1}{R} - \alpha \ . \tag{2.80}
$$

The wetted length of the wall in the channel cross-section is defined as:

$$
r_W = 
\begin{cases}
1 & ; \ k \geq k_c \ , \\[1.5ex]
R\cos(\alpha+\gamma) & ; \ k < k_c \ ,
\end{cases}
\tag{2.81}
$$

where $k_c$ is defined in equation (2.74).

The non-dimensional boundary conditions are:

$$
v(x=1) = v_1 = \left.\frac{Q}{A}\right|_{x=1} \ , \tag{2.82}
$$

$$
h(x=0) = h_0 \ , \tag{2.83}
$$

$$
k(x=0) = \Lambda \ , \tag{2.84}
$$

$$
k(x=1) = \Lambda \ . \tag{2.85}
$$

In addition, normal and tangential viscous stresses at the gas-liquid interface are assumed to be negligible (slip condition). In dimensionless form, the pressure boundary condition at the inlet (compare equation (2.45)) yields:

$$h_0 = K_0 + K_1 \frac{\text{Oh}v_0}{4} + K_2 \frac{v_0^2}{2} \quad . \tag{2.86}$$

The coefficients $K_0, K_1, K_2$ are identical to those in equation (2.45). Accordingly, the dimensionless boundary condition for two-phase flow yields:

$$h_0 = K_0 + K_1 \frac{\text{Oh}v_0}{4} + K_2 \frac{v_0^2}{2} + \frac{\Delta p'_{am}}{p_c} \quad . \tag{2.87}$$

# Chapter 3

# Experiments

This chapter presents the most important details of the capillary channel flow (CCF) experiment setup, materials, and methods. Experiment procedures and test matrices are also discussed. The results obtained from the experiments are presented in chapter 6.

## 3.1   Experiment Setup

The experimental hardware was manufactured by Airbus Defence and Space (formerly EADS Astrium) and was designed to operate within the microgravity science glovebox (MSG) onboard the ISS. Figure 3.1 shows how the individual components of the CCF setup are installed in the MSG. The interested reader will find additional information on the technical specifications of the MSG in Spivey et al. [57]. The experiment hardware consists of four modules: two experiment units (EU1 and EU2), the electrical subsystem (ESS), and the optical diagnostic unit (ODU). EU1 and EU2 both contain an entire fluid loop in a sealed container with windows for illumination and observation of the test channel. The main difference between EU2 and EU1 is the geometry of the respective test channels, which are shown in figures 3.3 and 3.4. The entire fluid loop, which differs slightly between the EUs, is shown

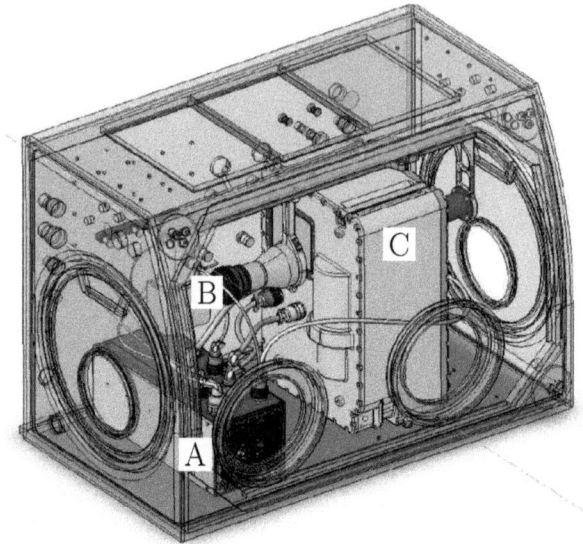

Figure 3.1: A rendered image of the CCF hardware components installed in the MSG onboard the ISS displaying the ESS (A), the ODU (B), and the EU (C).

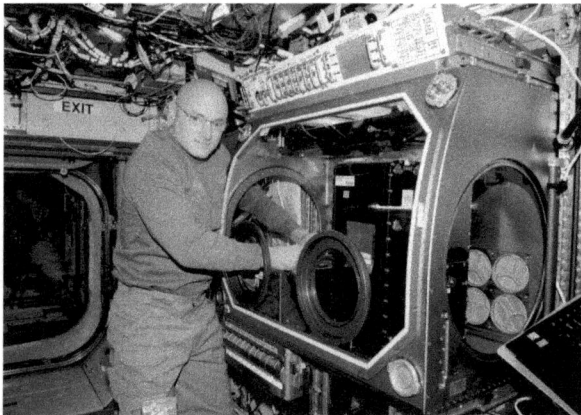

Figure 3.2: Commander Scott Kelly installing the CCF hardware into the MSG onboard the ISS (Source: NASA).

in figure 3.5. Each experiment setup consists of three modules. The first experiment setup consists of the ESS, ODU, and EU1. A member of the ISS crew is required to install the equipment in the MSG as shown in figure 3.2. For the second setup, EU1 is removed by a crew member and replaced by EU2. The ODU and ESS are compatible with both EUs and are used for both setups. The ESS contains the power supply and electronic control units. The ODU consists of a high speed camera and a parallel light source. The specifications of the individual components are listed in table 3.1.

In EU1 the test channel (TC) consists of parallel-plates (PP) with two free surfaces opposite each other. In EU2 the test channel consists of a wedge-shaped capillary channel (WE) with a triangular cross-section, which means that it has only one available free surface. Slide bars alter the length of the channel and, in EU1, can be used to open only one side of the test channel to form a rectangular groove-shaped channel (GR) with a single free surface. The test channel is located within the test unit (TU), which provides a closed gaseous environment. Furthermore, plunger K1 is present only in EU1 and the bubble injection system (BI) is installed only in EU2.

Experiments were performed around the clock via telemetric commanding from one of two ground stations located in Bremen, Germany and in Portland (OR), USA. Low rate telemetry data (one measurement per second) was recorded at both ground stations which included readings from 6 pressure sensors and 16 temperature sensors as well as data from the flow meter. A low resolution live video stream of the MSG cameras was also recorded. For selected experiments additional images were recorded using the high resolution high speed camera.

## 3.2  Fluid Loop

The main components of the experiment unit were tested extensively in ballistic rocket flights [44, 46] and drop tower experiments [22, 30, 33]. The fluid loop is shown schematically in figure 3.5. Following the direction of flow, fluid passes through

Table 3.1: Specifications of the main components of the experiment hardware.

| component | dry mass $m'/\text{kg}$ | width $x'/\text{m}$ | depth $y'/\text{m}$ | height $z'/\text{m}$ |
|---|---|---|---|---|
| EU1 | 37.2 | 0.326 | 0.480 | 0.348 |
| EU2 | 36.9 | 0.326 | 0.480 | 0.348 |
| ESS | 8.5 | 0.414 | 0.178 | 0.170 |
| ODU | 5.4 | 0.125 | 0.257 | 0.768 |

the test channel (TC), valve C4, the pump (P), the flow meter (F), the phase separation chamber (PSC), the flow preparation chamber (FPC), valve C9, and then re-enters the test channel. A flow chart showing additional components such as sensors for temperature and pressure is displayed in appendix B. The length of the test channel is the length of the free surface which is defined by the distance between the pinning edges of the inlet and outlet of the test channel. The pinning edge at the inlet of the test channel is stationary and points downstream. The pinning edge at the outlet of the test channel is located on the end of the sliders and points upstream. The thickness of each pinning edge is $0.2 \times 10^{-3}\,\text{m}$. The sliders (S1, S2) can be moved individually to vary the length of the test channel within the range $0.1 \times 10^{-3}\,\text{m} \le l' \le 48 \times 10^{-3}\,\text{m}$. The slider actuators are piezo linear motors that allow a high degree of accuracy ($\pm 0.05 \times 10^{-3}\,\text{m}$).

The test channels are made of quartz glass. In EU1, the distance between the parallel glass plates is $a = 5 \times 10^{-3}\,\text{m}$ and the height of the channel is $b = 25 \times 10^{-3}\,\text{m}$ (compare figure 3.4). In EU2, the test channel opening angle of the triangular cross-section is $2\alpha = 15.8°$ (see figure 3.3). The height of the test channel is $b = 30 \times 10^{-3}\,\text{m}$ and the distance between the tilted faces at the top of the channel is $a = 2b\tan\alpha = 8.32 \times 10^{-3}\,\text{m}$. The cross-sectional area of the inlet of both test channels is $A_0' = 1.25 \times 10^{-4}\,\text{m}^2$.

The geometry of the fluid loop in the vicinity of the test channel of EU2 is shown in figures 3.5 and 3.6. The total cross-sectional area of the FPC is $7.238 \times 10^{-3}\,\text{m}^2$

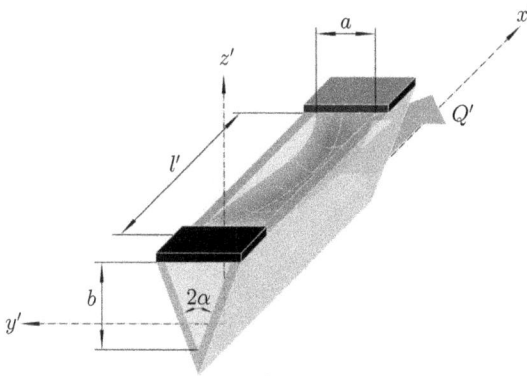

Figure 3.3: Liquid flowing through a triangular wedge that forms an open capillary channel. The geometry of the channel is defined by its length $l'$, its opening angle $2\alpha$, and its height $b$. A gas injection needle (not shown here) is located above the vertex at the inlet of the channel.

and flow passes along deflectors (D) and through a perforated sheet (PS) to enter the FPC with a velocity distribution that is assumed to be uniform. The sum of the perforations in the PS reduces the effective inlet area to $A'_{FPC} = 1.018 \times 10^{-3}\,\mathrm{m}^2$ through which flow is possible. Upstream of the test channel, the nozzle is $s'_1 = 30 \times 10^{-3}\,\mathrm{m}$ long and has an elliptical contour in the $x'y'$-plane in EU1 and in both planes in EU2. The cross-sectional area of the nozzle's inlet is $A'_{N1} = 0.75 \times 10^{-3}\,\mathrm{m}^2$ for EU1 and $A'_{N2} = 0.609 \times 10^{-3}\,\mathrm{m}^2$ for EU2. At its outlet, the cross-section of the nozzle is identical to $A'_0$. The nozzle is connected to the test channel by a closed entrance duct with a cross-section that is identical to $A'_0$. The length of the entrance duct is $s'_2 = 33.3 \times 10^{-3}\,\mathrm{m}$. As explained above, the length of the test channel is variable with $s'_3 = l'$. Downstream of the test channel, liquid passes through an exit duct which initially has the same cross-section as the test channel itself. The length of the exit duct depends on the length of the test channel such that $s'_4 = 67 \times 10^{-3}\,\mathrm{m} - l'$. The cross-section then constricts over a length of $s'_5 = 12.8 \times 10^{-3}\,\mathrm{m}$ to a circular cross-section with a diameter of $10.7 \times 10^{-3}\,\mathrm{m}$.

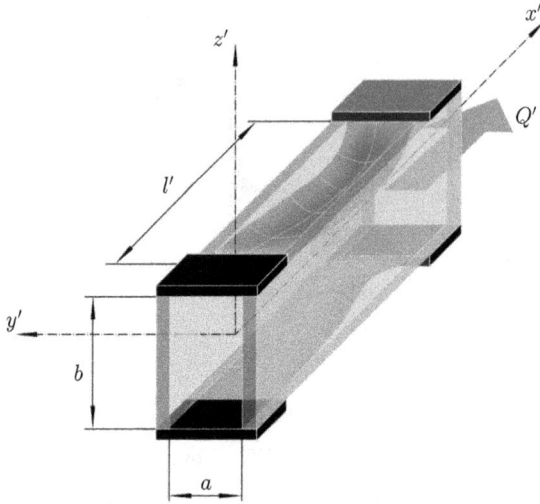

Figure 3.4: Liquid flowing through parallel plates that form an open capillary channel. The geometry of the channel is defined by its length $l'$, its width $a$, and its height $b$.

The compensation tube (CT, figure 3.5) is cylindrical and has a diameter of $2R'_{CT} = 60 \times 10^{-3}$ m and a height of $100 \times 10^{-3}$ m. Its cylindrical wall is transparent so that the fill level may be recorded during operations with a camera located in the MSG workspace. The compensation tube provides an additional free surface (FS) in the system. This liquid meniscus defines the static pressure at the inlet of the test channel. The compensation tube also compensates varying volumes during unsteady flow. For example, when gas is ingested or injected into the liquid loop, liquid is displaced and the fill level in the compensation tube rises without changing the pressure boundary conditions of the test channel. The liquid in the compensation tube is connected to the liquid in the flow preparation chamber by a tube that is $97.7 \times 10^{-3}$ m long with an inner diameter of $7.75 \times 10^{-3}$ m that bends by $90°$. The gas in the compensation tube is connected to the gas environment in the TU. This connection may be closed using valve C11.

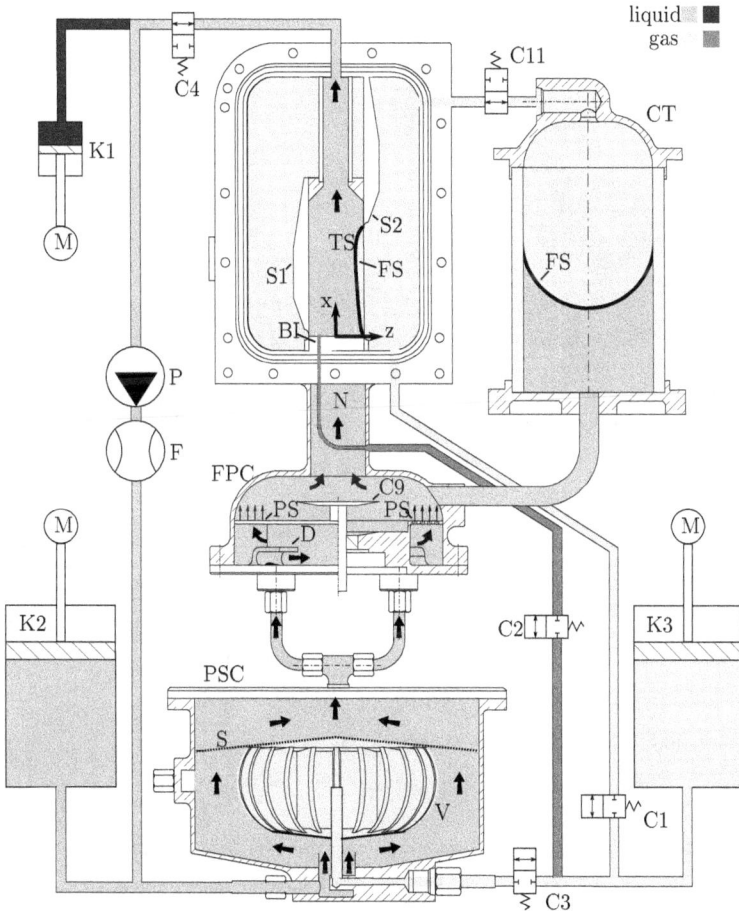

Figure 3.5: Schematic of the fluid loop inside each EU. The fluid management system contains liquid (cyan and ultramarine) and gas (yellow and green). Ultramarine lines exist only in EU1 and green lines exist only in EU2. The main components of the fluid loop are described in section 3.2.

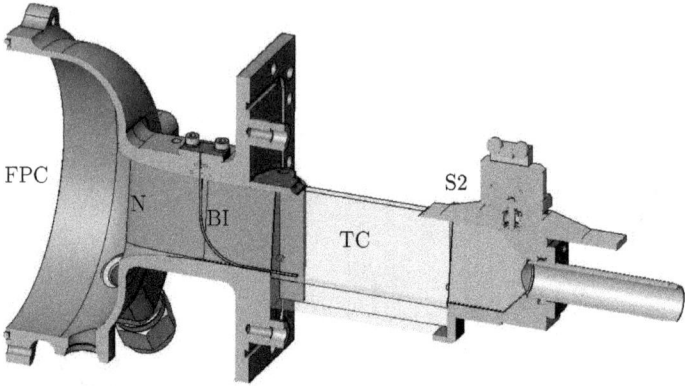

Figure 3.6: The region of interest in the flow path of the experiment setup extends from the FPC upstream of the TC to the circular exit duct downstream of the TC. The rendering is cut in the symmetry plane ($xz$-plane) of the TC. The components of the fluid loop are described in section 3.2.

The forced flow is maintained by a Micropump GB-P25 pump head and a Maxon EC32 brushless motor. In the experiments the flow rate $Q'$ is limited to a maximum flow rate of $Q' \leq 20 \times 10^{-6}\,\mathrm{m^3\,s^{-1}}$. Both flow rate and flow acceleration can be varied in the experiments. The flow meter is a DPM-1520-G2 turbine flow meter from Kobold and is calibrated to an accuracy of $\pm 0.1 \times 10^{-6}\,\mathrm{m^3\,s^{-1}}$. Flow is considered to be laminar throughout the experiments as Reynolds numbers[1] are consistently lower than 1800 at the inlet of the channel.

K2 and K3 are plunger bellows and are the reservoirs for the liquid and the gas phases, respectively. Both bellows have an operational volume of $273 \times 10^{-6}\,\mathrm{m^3}$. Between experiments, the plungers are used for fluid management and setting boundary conditions including the fill level within the CT, the gas volume within the PSC, and the pressure within the gaseous environment surrounding the test channel. When choking occurs these parameters change during the experiment because the ingested

---

[1]$\mathrm{Re} = \bar{v}'_0 d'_h \nu^{-1}$, where $\bar{v}'_0$ is the mean velocity at the inlet of the TC and $\nu$ is the kinematic viscosity of the liquid. $d'_h$ is the hydraulic diameter with $d'_h = 4A'_0/P'$ and $P'$ is the wetted perimeter.

gas is collected in the PSC and displaces volume in the CT. Plunger K1 can impose an oscillation on the flow rate at the outlet of the test channels in EU1 with volume amplitudes of up to $\pm 0.5 \times 10^{-6}\,\mathrm{m}^3$ and frequencies up to $10\,\mathrm{s}^{-1}$.

In EU2 a bubble injection system is integrated within the fluid loop. A cannula penetrates the wall of the entrance duct opposite the vertex of the channel and bends around to ultimately point in flow direction. The outlet of the cannula is located immediately upstream of the inlet of the test channel ($x' = 0$) and approximately $3.6 \times 10^{-3}\,\mathrm{m}$ above the vertex of the wedge (see figure 3.3). The cannula has an inner diameter of $0.6 \times 10^{-3}\,\mathrm{m}$ and an outer diameter of $0.8 \times 10^{-3}\,\mathrm{m}$. Gas injection frequency and bubble size can be altered by defining the operating parameters of C2 which is a solenoid valve of model MHE2-MS1H-3/2O-M7 from Festo and connects the cannula to the gas reservoir in plunger K3. The frequency of valve C2, $f'_{C2}$, can be varied in the range of 0 to $10\,\mathrm{s}^{-1}$ (i.e. gas line closed to opening once per 0.1 s). The duty cycle, $\Delta t'_{dc}$, of valve C2 defines how long the gas line is open and can be varied in the range of $0\,\mathrm{s} \leq \Delta t'_{dc} \leq 10\,\mathrm{s}$.

The PSC is designed to gather any gaseous phase which may have entered the fluid loop and prevent it from re-entering the test channel. Gas is trapped in the PSC by a porous screen (S, figure 3.5) that has a nominal pore size of $5 \times 10^{-6}\,\mathrm{m}$. A bubble shaped vane structure (V, figure 3.5) ensures that gas is kept in the centre of the PSC. The size of the bubble in the phase separation chamber is determined qualitatively using three 10K3MCD1 Betatherm NTC thermistors which have an accuracy of $\pm 0.2\,^{\circ}\mathrm{C}$ at $25\,^{\circ}\mathrm{C}$. Dissipation of the heat that is generated by the thermistors depends on the properties of the fluid that it is immersed in. In this way local temperature measurements indicate the presence of the gaseous or liquid phase. The three thermistors are located along an axis as indicated in figure 3.7. Based on Surface Evolver [7] computations the bubble volume at bubble sensors 1, 2, and 3 is estimated to be $90 \times 10^{-6}\,\mathrm{m}^3$, $150 \times 10^{-6}\,\mathrm{m}^3$, and $200 \times 10^{-6}\,\mathrm{m}^3$, respectively. During the experiments the interface of the gas bubble in the PSC was located between bubble sensor 1 and bubble sensor 3. Gas can be extracted from the PSC through a needle that protrudes into its centre (see figure 3.5). The needle

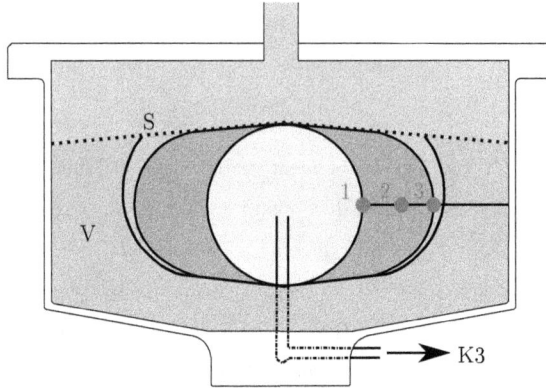

Figure 3.7: The bubble sensor system in the PSC. During the experiments, the interface of the bubble in the PSC was always located between sensor 1 (yellow bubble) and sensor 3 (green bubble). The test liquid is coloured blue and the vanes (V) keeping the bubble in the centre are grey. The porous screen is displayed as a dotted line.

is connected to plunger K3 via valve C3. As already discussed, re-entry of the liquid flow into the test channel occurs after passing through the FPC where a uniform laminar flow is developed by means of deflector plates, a perforated sheet, and the nozzle.

The test fluid that was used in the experiments is the commercially available 3M$^{TM}$ Novec$^{TM}$ 7500 Engineered Fluid, which is a hydrofluoroether that is generally used for heat transfer and for cleaning electronic parts. The properties of the test liquid are listed in table 3.2 for the temperature range that is relevant for the experiments. The total volume of test liquid in the system is $1.68 \times 10^{-3} \, \text{m}^3$. The gaseous environment consists of nitrogen. During operations, the system pressure was typically in the range of $0.95 \times 10^5$ Pa to $1.15 \times 10^5$ Pa. The temperature dependency of the physical properties of the test liquid was determined beforehand and yielded following equations with temperature $T'$ in °C:

Table 3.2: Density $\rho$, viscosity $\nu$, and surface tension $\sigma$ of the test liquid Novec$^{TM}$ Engineered Fluid HFE-7500 at different temperatures $T'$. Its static contact angle on quartz glass is $\gamma = 0°$.

| $T'/°C$ | $\rho_L/(\mathrm{kg\,m^{-3}})$ | $\nu_L/(10^{-6}\,\mathrm{m^2\,s^{-1}})$ | $\sigma/(10^{-3}\,\mathrm{kg\,s^{-2}})$ |
|---|---|---|---|
| 25.0 | $1620 \pm 2$ | $0.773 \pm 0.1$ | $16.66 \pm 0.1$ |
| 30.0 | $1610 \pm 2$ | $0.715 \pm 0.1$ | $16.16 \pm 0.1$ |
| 35.0 | $1600 \pm 2$ | $0.664 \pm 0.1$ | $15.67 \pm 0.1$ |

$$\rho_L/(\mathrm{kg\,m^{-3}}) = 1670.9833 - 2.02T' \quad , \tag{3.1}$$
$$\nu_L/(10^{-6}\,\mathrm{m^2\,s^{-1}}) = \exp\left(-3.2319 + \frac{555.62}{T' + 161.83}\right) \times 10^{-6} \quad , \tag{3.2}$$
$$\sigma/(10^{-3}\,\mathrm{kg\,s^{-2}}) = (19.12025 - 0.09863T') \times 10^{-3} \quad . \tag{3.3}$$

The temperature within the experimental setup is monitored at various locations within the liquid and gas loops using PT1000 resistance temperature detectors from Honeywell with an accuracy of $\pm 0.5\,°C$ within the temperature range that was used in the experiments. The pressure is monitored in the gas duct near plunger K3, in the liquid within the FPC, and in the gas environment surrounding the test channel. The pressure sensors are PAA-33X models from Keller, which are calibrated to an absolute accuracy of less than $100\,\mathrm{Pa}$. Throughout the experiments the temperature remained within the range of $25\,°C$ to $35\,°C$. The average temperature during operations with EU1 was $30.4\,°C$ and during operations with EU2 it was $29.8\,°C$. According to the observed temperature range and following equations (3.1), (3.2), and (3.3), the Ohnesorge number range yields $2.12 \times 10^{-3} \leq \mathrm{Oh} \leq 2.41 \times 10^{-3}$ and $2.37 \times 10^{-3} \leq \mathrm{Oh} \leq 2.65 \times 10^{-3}$ for EU1 and EU2 respectively[2]. The respective average Ohnesorge numbers are determined at $T' = 30\,°C$ and yield $\mathrm{Oh} = 2.26 \times 10^{-3}$ and $\mathrm{Oh} = 2.49 \times 10^{-3}$ for EU1 and EU2 respectively. Cooling of the experiment

---

[2]Ohnesorge numbers are determined in accordance with equation (2.63) and only parallel plates are considered here for EU1 with $d_h = 2a$.

apparatus is performed by the MSG work volume air circulation system and air handling unit, which provide 200 W of cooling power [57]. In addition, each EU is equipped with a thermal control system (TCS) with thermoelectric coolers, which transfer heat from the liquid loop to the air in the MSG work volume.

### 3.2.1    Flow Rate Conversion

A comparison of the flow meter readings and the flow rate setting for the pump was performed for EU2.1 and for EU2.2[3]. Using this data, the defined pump setting $Q'_P$ was converted into the more accurate flow meter reading $Q'_{FM}$ using the following equation for EU2.1( in units $10^{-6} \, \mathrm{m}^3 \, \mathrm{s}^{-1}$):

$$Q'_{FM} = 0.987 \, Q'_P - 0.11 \quad . \tag{3.4}$$

The conversion equation for EU2.2 was found to be ( in units $10^{-6} \, \mathrm{m}^3 \, \mathrm{s}^{-1}$):

$$Q'_{FM} = 0.997 \, Q'_P - 0.08 \quad . \tag{3.5}$$

The results of the calibration measurements used for equations (3.4) and (3.5) are shown in figure 3.8. This linear calibration function was used to calculate a more accurate flow rate when the flow sensor could not be used to determine the actual flow rate. This is especially the case in two-phase flow experiments where the gaseous phase affects the readings of the flow sensor that is calibrated for single-phase flow only. In the flow rate range of the experiments, the maximum deviation between the conversion equations and the flow meter readings was found to be $0.0124 \, Q'_{FM}$ for equation (3.4) and $0.013 \, Q'_{FM}$ for equation  (3.5).

---

[3]EU2 was installed multiple times and the respective experiment phases are abbreviated with EU2.1, EU2.2, etc. The order in which the experiment phases occurred is described by the digit after the dot.

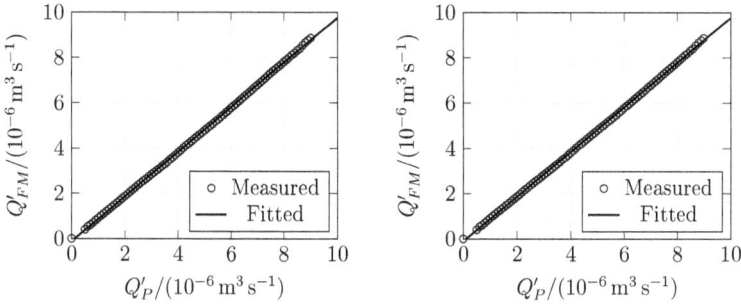

Figure 3.8: Calibration fits for flow rate conversion between pump setting and flow meter reading. The control value $Q'_P$ is the set flow rate of the pump. The indicated value $Q'_{FM}$ is the observed flow meter reading. The linear calibration functions were fitted to the measurements using equation (3.4) (left) and equation (3.5) (right).

## 3.3  Optical Setup

The ODU consists of a high speed high resolution camera (HSHRC) and a parallel light source that are positioned perpendicularly to the outer walls of the test channel (see figure 3.9) thus recording in the $x'z'$-plane of the respective test channels (compare figure 3.4 and 3.3). The camera is a Motion BLITZ Cube 26H from Mikrotron, which can record up to 250 frames per second at a resolution of 1280 x 1024 pixels per frame. A large-field telecentric lens is attached to the camera. The test channel is illuminated by a Correctal TC parallel light source from Sill Optics. The LEDs emit light at a wavelength of $\lambda = 455 \times 10^{-9}$ m, which corresponds to the optimal wavelength sensitivity of the CCD chip in the camera. The parallel light source and the HSHRC are mounted on an optical bench to ensure they are aligned along an axis. The HSHRC is set up to capture the entire test channel. The optical resolution of the test sections is computed to be $0.038 \times 10^{-3}$ m/pixel. The field of view contains the test channel and covers an area of $49.3 \times 10^{-3}$ m by $39.4 \times 10^{-3}$ m. Recorded data is processed onboard and then transferred to the MSG Laptop Computer (MLC) and subsequently to the ground stations. The onboard

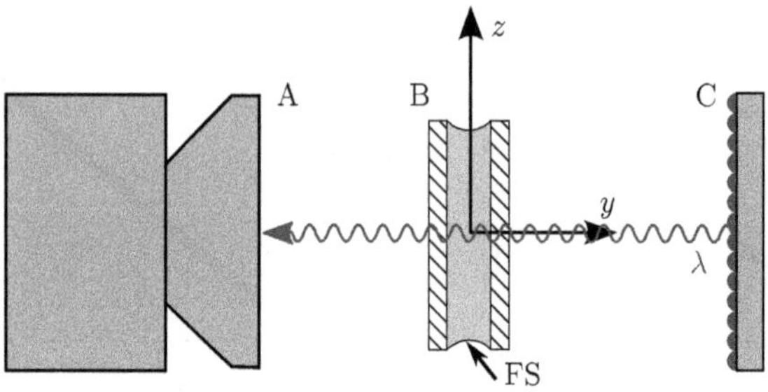

Figure 3.9: Optical setup for the HSHRC (not to scale) with the camera (A), the test channel (B), and the light source (C). Light that is refracted at the free surfaces (FS) appears in recorded images as a black area with a well defined edge against the illuminated background.

image processing software contains of an edge detection algorithm and converts the recorded 8 bit monochrome images to binary images. Onboard image processing reduces the file size considerably and increases download efficiency.

Further image analysis is performed later to determine the contour of the free surface in the test channel based on the processed black and white images. The combination of onboard image processing and subsequent analysis of the contour leads to an error of approximately 2 to 3 pixels, which corresponds to less than $0.12 \times 10^{-3}$ m.

Additional Hitachi HV-C20 CCD cameras that belong to the inventory of the MSG were installed to observe the CT (MSG camera 1) and the test channel (MSG camera 2). Throughout operations one ISS video channel was dedicated to broadcasting a live stream of one of the MSG cameras. A live feed of one of the cameras was broadcast to the ground stations at a frame rate of at least 8 fps. Throughout all experiments, video data was recorded from the live stream obtained from one of the MSG video cameras.

## 3.4 Electrical Subsystem

The ESS contains all required electronic boards including the motion controllers and the CCF computer. The CCF computer is a MOPS-PM from KONTRON with 512 MB RAM and a 1 GHz Intel Celeron M processor. Two 512 MB compact flash memory cards are used as hard-drives. The ESS connects to the MSG for power supply and for data transfer. Furthermore, an ethernet connection is established to the MSG laptop computer and is dedicated to downloading science data from the CCF computer. One of the main tasks of the CCF computer is to convert the grey-scale images recorded by the HSHR camera into black and white images to reduce the data volume. At all times, the ESS requires less than 200 W to power the entire experiment setup. Therefore, the total power consumption at any given time is less than the heat exchange capability of the MSG work volume air handling unit mentioned in section 3.2.

## 3.5 Initial Filling Procedure

During transportation and installation of the EU all valves in the fluid loop are closed, the CT and PSC are completely filled with the test liquid, and the line between C9 and C4 contains only gas. Before nominal operations can commence, the following procedures must be performed:

a) a gas-liquid interface must be established in the CT,

b) a bubble must be injected into the PSC,

c) the test channel must be filled with the test liquid.

Generating a gas-liquid interface in the CT must be accomplished first so that liquid that is displaced by injecting gas into the PSC can be compensated. This is accomplished by opening valve C11 and then pulling liquid into reservoir K2.

Care must be taken to prevent liquid from exiting the CT through C11 when it is opened. This can be prevented by setting the pressure of the gas in the TU slightly higher than that of the liquid in the CT by opening C1 and altering the gas volume in reservoir K3. Approximately 75 % of the liquid in the CT is transferred to the initally almost empty reservoir K2.

Next, a bubble is introduced into the PSC. As mentioned above, gas can be transferred from the gas reservoir K3 to the PSC and vice versa. The needle that protrudes into the centre of the PSC is connected to K3 via valve C3 (see figure 3.5). A sufficiently large bubble is required to ensure that the needle is located well within the bubble. Later, when gas has been ingested into the liquid loop due to choking, it can be retrieved from the PSC without also extracting liquid and subsequently gas may be relocated into the test unit, thereby restoring the ambient pressure boundary condition. Before opening valve C3, care must be taken that the pressure in the gas reservoir is higher than that of the liquid in the PSC thus avoiding flow of liquid into the gas lines. Gas is then injected stepwise into the PSC until the bubble interface has reached bubble sensor 1 and its size is approximately equal to that of the yellow region in figure 3.7. Subsequently, all valves are closed and the procedure for filling the TC can be performed.

The procedure for filling the TC differs between EU1 and EU2. In EU1 the TC is filled from both sides sequentially. With the slider opened to $34.4 \times 10^{-3}$ m, valve C4 and C11 are opened and then liquid is transferred from K2 through C4 into the TC until the advancing meniscus is observed at the outlet of the TC. Then C4 and C11 are closed again. Once all valves are closed, C9 is opened and liquid is transferred from K2 through the inlet of the TC until the meniscus at the outlet coalesces with the interface advancing from the inlet. Then C11 is opened again and a low flow rate is established with the pump.

In EU2 the TC is filled only from the outlet. First the slider is moved to open the TC by $20 \times 10^{-3}$ m. Then valve C4 is opened and liquid is transferred from K2 into the TC. The amount of liquid in the TC is increased in small steps and corner flow at the vertex of the wedge transports a portion of the liquid towards the inlet of the TC and C9. This is continued until the TC is filled from both sides and the gas in the TC has been expelled completely. Then valves C9 and C11 are opened and a low flow rate is established by the pump.

Once the filling procedure is completed, nominal operations commence and science data can be gathered. The filling procedure is performed via tele-commanding from the ground stations. For each EU the entire initial filling procedure was completed within 8 hours.

# 3.6   Experiment Procedures

## 3.6.1   Determining the Single-Phase Critical Flow Rate

In the steady flow experiments the critical flow rate[4], $Q'_{crit,1P}$, is defined as the maximum constant flow rate before gas ingestion is observed across the free surface. The single-phase critical flow rate, $Q'_{crit,1P}$ is determined by a step-wise approach for constant channel lengths in the range of 0.5 mm to 48 mm. Beginning at a subcritical flow rate at which the free surface is stable, the flow rate is increased by $0.01 \times 10^{-6} \, \mathrm{m^3 \, s^{-1}}$ over a time interval of 1 s. The stability of the interface is observed at a constant flow rate using the live video stream for at least 30 s. If the contour of the free surface appears stable then the flow rate is increased further. This process of incrementally increasing the flow rate and subsequent observation is repeated until gas ingestion is observed. Gas ingestion across the interface signals that the flow rate has exceeded the critical flow rate and the channel is choked. A typical time-line of the flow rate is displayed in figure 3.10. Once gas ingestion occurs, the interface is restabilized by decreasing the flow rate by a sufficient amount. Following the restabilization of the interface, $Q'_{crit,1P}$ is carefully established once

---

[4]In most cases, the indices explicitly display whether a flow rate is considered to be single-phase or two-phase. In some cases this index is withheld for better readability, but the respective case will be obvious from the context.

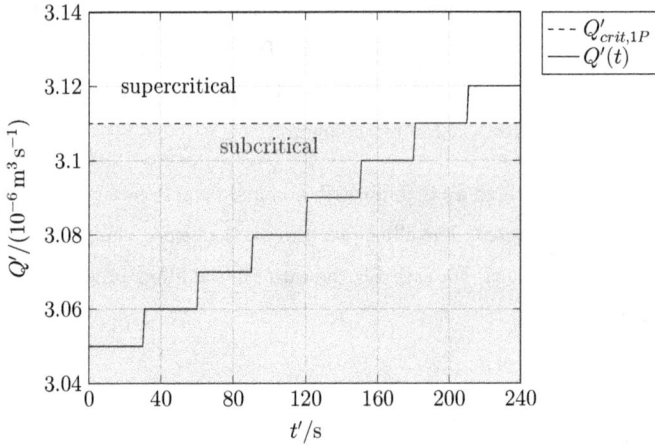

Figure 3.10: Typical flow rate evolution $Q'(t')$ during an experiment to determine the single-phase critical flow rate $Q'_{crit,1P}$. The flow rate is increased incrementally until choking occurs during an interval of constant $Q'$ (here at $3.12 \times 10^{-6}\,\mathrm{m^3\,s^{-1}}$). $Q'_{crit,1P}$ is defined as the highest flow rate at which a stable interface is observed during an interval with constant $Q'$.

more for video data acquisition with the HSHRC. The duration of the recordings are variable and are typically selected to be between 2 seconds and 20 seconds depending on the selected frame rate. Once the recording is finished, onboard image processing is initiated, which typically takes approximately 30 minutes to complete. Additional data recorded from each experiment include the channel geometry (shape and length), pressure of the liquid in the FPC (sensor[5] PressureC9) and ambient gas pressure (sensor Pressure1SW), average temperature in the TC (mean value of sensors EUTemp5SW and EUTemp6SW), flow meter reading, pump speed setting, and the time of the experiment. If HSHRC images are recorded successfully, then the flow rate is determined by averaging the flow rate reading that is found in the

---

[5]An overview of the most important sensors and their locations in the fluid loop is located in appendix B.

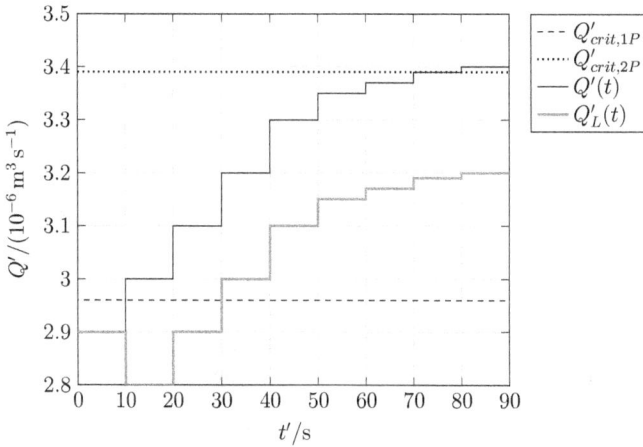

Figure 3.11: Generic flow rate evolution $Q'(t') = Q'_G + Q'_L(t)$ during an experiment to determine the two-phase critical flow rate $Q_{crit,2P}$. Single-phase critical flow rate $Q'_{crit,1P}$ is displayed for reference. Gas injection is initiated at $t' = 10\,\mathrm{s}$ with $f'_{C2} = 2\,\mathrm{s}^{-1}$ and with time-averaged gas injection flow rate $Q'_G = 0.2 \times 10^{-6}\,\mathrm{m}^3\,\mathrm{s}^{-1}$. The pump speed is increased incrementally until choking occurs at $Q' = 3.40 \times 10^{-6}\,\mathrm{m}^3\,\mathrm{s}^{-1}$. $Q'_{crit,2P}$ is defined as the highest flow rate at which a stable interface is observed during an observation interval with constant $Q'$, in this case at $Q'_{crit,2P} = 3.39 \times 10^{-6}\,\mathrm{m}^3\,\mathrm{s}^{-1}$.

header data of each individual frame image. On average, each individual data point for determining $Q'_{crit,1P}$ is repeated approximately three times.

## 3.6.2  Determining the Two-Phase Critical Flow Rate

The two-phase critical flow rate, $Q'_{crit,2P}$, is determined by a step-wise approach for a constant channel length in the range of 5 mm to 30 mm. Gas is injected into the flow immediately upstream of the inlet of the test channel. The flow rate at the outlet of the channel is defined as $Q' = Q'_G + Q'_L$, where $Q'_L$ is the flow rate of the liquid and $Q'_G$ is the time-averaged flow rate of the injected gas:

$$Q'_G = f'_{C2} V'_{C2} \quad , \tag{3.6}$$

with the bubble injection frequency $f'_{C2}$ and the average volume of the injected bubbles $V'_{C2}$. $Q'_G$ and $Q'$ are controlled parameters that are respectively set by defining the parameters of the bubble injector (duty cycle, frequency, and pressure difference DiffPressureC9) or by varying the pump speed. The average gas flow rate injected into the channel is adjusted via the pressure difference between the gas reservoir K3 and $p'_0$ in accordance with Weislogel et al. [64]. $Q'_L$, on the other hand, depends on the set values of $Q'_G$ and $Q'$ such that $Q'_L = Q' - Q'_G$. For $Q'_G > 0$, liquid is displaced from the FPC into the CT at a flow rate equal to $Q'_G$. Therefore the total flow rate in the nozzle section of the flow loop up to the inlet of the test channel is $Q'_L$.

At the beginning of an experiment, flow is single-phase ($Q' = Q'_L$) with $Q'$ just below $Q'_{crit,1P}$. Periodic gas injection is initiated using the bubble injector with a pre-determined injection frequency of $f'_{C2} = 2\,\mathrm{s}^{-1}$, duty cycle of $\Delta t'_{C2} = 0.2\,\mathrm{s}$ and bubble volumes in the order of $0.05 \times 10^{-6}\,\mathrm{m}^3\,\mathrm{s}^{-1} \leq V'_{C2} \leq 0.2 \times 10^{-6}\,\mathrm{m}^3\,\mathrm{s}^{-1}$ (*ergo* a time-averaged constant gas flow rate of $0.1 \times 10^{-6}\,\mathrm{m}^3\,\mathrm{s}^{-1} \leq Q'_G \leq 0.4 \times 10^{-6}\,\mathrm{m}^3\,\mathrm{s}^{-1}$). Gas injection generates a train of mono-disperse bubbles that are transported through the channel. The bubbles assume a spherical shape within the first few millimetres of the test channel and continue to be injected throughout the experiment. Immediately after the initiation of gas injection, $Q'_{crit,2P}$ is approached by incrementally increasing the total flow rate $Q'$ within the channel in steps of up to $0.1 \times 10^{-6}\,\mathrm{m}^3\,\mathrm{s}^{-1}$ over a time interval of $1\,\mathrm{s}$ per step increase.

The stability of the interface is observed at each constant flow rate interval using the live video stream for approximately $10\,\mathrm{s}$. If the contour of the free surface appears stable then the flow rate is increased further. This process of incrementally increasing the flow rate with subsequent observation is repeated until gas ingestion is observed. Gas ingestion across the interface signals that the critical flow rate has been exceeded and that the channel is choked. A typical time-line of the flow rate is displayed in figure 3.11. Throughout the two-phase experiments, the flow meter signal is disturbed by gas bubbles flowing through the sensor. The flow rate must therefore be calculated from the pump speed setting as described in section 3.2.1.

### 3.6.3 Steady Subcritical Flow Experiments

Subcritical flow experiments were performed after having determined $Q'_{crit,1P}$. The experiments were performed only for single-phase flow using constant flow rates $0.8\,Q'_{crit,1P}$ and $0.9\,Q'_{crit,1P}$ at constant channel lengths $5 \times 10^{-3}\,\mathrm{m} \leq l' \leq 45 \times 10^{-3}\,\mathrm{m}$. The HSHRC was used to record video data for later evaluation of the interface coordinates.

### 3.6.4 Steady Supercritical Flow Experiments

Supercritical flow experiments were performed for single-phase flow after having determined $Q'_{crit,1P}$. Experiments were performed using constant flow rates from $0.25 \times 10^{-6}\,\mathrm{m^3\,s^{-1}}$ to $2.5 \times 10^{-6}\,\mathrm{m^3\,s^{-1}}$ above $Q'_{crit,1P}$ at various constant channel lengths $4 \times 10^{-3}\,\mathrm{m} \leq l' \leq 20 \times 10^{-3}\,\mathrm{m}$. The HSHRC was used to record video data for later evaluation of the volume of the ingested bubbles and the bubble ingestion frequency.

## 3.7 Experiment Test Matrix

A total number of 1021 experiments was recorded for single-phase steady state flow. The test matrix for data acquisition of the single-phase experiments is summarized in figure 3.12. The test matrix was defined dynamically by first determining $Q'_{crit,1P}$ and then subsequently defining additional data points in the subcritical and supercritical regimes. For the majority of examined $l'$, data points were repeated and values for $Q'_{crit,1P}$ were averaged.

The number of two-phase experiments that are discussed in this work amounts to 26. The two-phase experiments were performed after having determined the single-phase critical flow rate and after the supercritical single-phase experiments. The test matrix for the discussed two-phase experiments is displayed in figure 3.13.

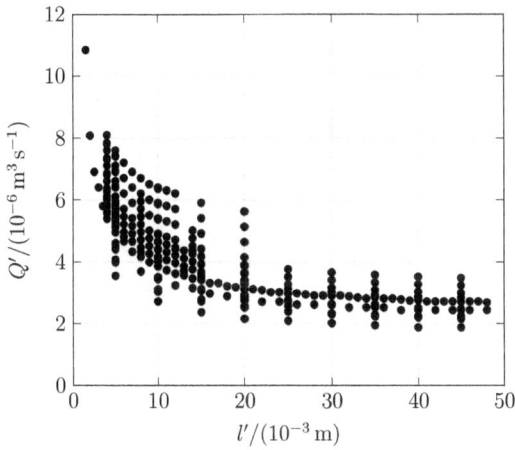

Figure 3.12: Test matrix for all single-phase steady state flow experiments. Many test cases were repeated and therefore individual data points may represent multiple experiment runs.

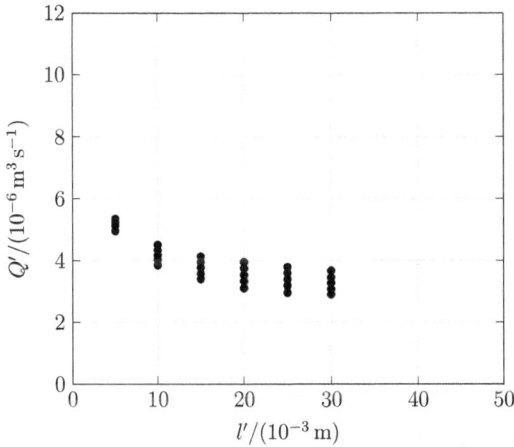

Figure 3.13: Test matrix for the discussed two-phase flow experiments.

# Chapter 4

# Three-Dimensional Numerics

Three-dimensional numerical simulations of the presented flow problem were performed with the CFD toolbox OpenFOAM (version 2.2.x) [38, 65]. The advantage
of OpenFOAM lies in its Open-Source licensing under GPL (General Public Li-
cense) which means that users are free to inspect, use, and modify the source code.
OpenFOAM uses the Finite Volume Method (FVM) to discretize a continuous computational domain. Single-phase simulations were performed with OpenFOAM's
solver simpleFoam to determine pressure loss within the flow path upstream of the
test channel. The results were used to approximate the pressure at the inlet of the
test channel as a function of the Reynolds number. The fitted function was then
implemented in the 1D model as an inlet boundary condition. SimpleFoam is based
on the SIMPLE method [25, 35, 36] (Semi Implicit Method for Pressure Linked
Equations) and is applicable to single-phase, steady state, incompressible, isothermal flow problems within a fixed grid. Two-phase simulations were performed with
interFoam to determine $Q'_{crit,3D}$ and interface contours for subcritical flow. The ap-
plicability of the respective solvers to the computed flow problems was determined
via preliminary numerical studies on pressure loss in pipe flow and capillary rise at
a wall. These test cases are outlined in appendices D and E.

The interFoam solver is based on the two-fluid model of multiphase flow and
applies the volume of fluid (VOF) method to solve for isothermal, transient, two-

phase flow of two immiscible, incompressible fluids within a fixed grid [48]. A scalar
volume fraction field is used to determine the volume fraction $\beta$ of each discretized
cell where

$$\beta(x', y', z', t') = \begin{cases} 0 & \text{if occupied by fluid 0 (G),} \\ 0 < \beta < 1 & \text{at the interface,} \\ 1 & \text{if occupied by fluid 1 (L).} \end{cases} \tag{4.1}$$

In this thesis, fluid 0 is the ambient gas (G) and fluid 1 is the test liquid (L). The
interface between the two phases is not sharp and smeared over a thin band of cells
with $0 < \beta < 1$ in which the material properties are determined by weighting the
individual material properties of the two phases according to the phase fraction:

$$\rho = \beta \rho_L + (1 - \beta) \rho_G \quad , \tag{4.2}$$

$$\mu = \beta \mu_L + (1 - \beta) \mu_G \quad . \tag{4.3}$$

The VOF model numerically solves the equations for conservation of mass and momentum for both phases with indices $i = \{1, 2, 3\}$ and $j = \{1, 2, 3\}$ indicating the
tensor dimension:

$$\frac{\partial v_i'}{\partial x_i'} = 0 \quad , \tag{4.4}$$

$$\rho \left( \frac{\partial v_i'}{\partial t'} + v_j' \frac{\partial v_i'}{\partial x_j'} \right) = -\frac{\partial p'}{\partial x_i'} + \frac{\partial \tau_{ji}'}{\partial x_j'} + \rho g_i' + F_\sigma' \quad , \tag{4.5}$$

where $\tau_{ij}'$ is the stress tensor. A mass velocity $v_i'$ is defined as

$$v_i' = \frac{[\beta \rho_L v_{i,L}' + (1 - \beta) \rho_G v_{i,G}']}{\rho} \quad . \tag{4.6}$$

The volume fraction field is advected by the velocity field using the transport equation

$$\frac{\partial \beta}{\partial t'} + \frac{(\partial \beta u_i')}{\partial x_i'} + \frac{\partial}{\partial x_i'} \left[ \beta(1 - \beta) u_{i,\beta}' \right] = 0 \quad , \tag{4.7}$$

where $u_i'$ is the volumetric flux:

$$u_i' = \beta v_{i,L}' + (1 - \beta) v_{i,G}' \quad , \tag{4.8}$$

and $u_{i,\beta}'$ is the slip velocity between the phases:

$$u_{i,\beta}' = v_{i,L}' - v_{i,G}' \quad . \tag{4.9}$$

In contrast to the conventional VOF method, interFoam implements an interface compression method to reduce the numerical smearing of the interface between the two phases [4] by including an additional term to the transport equation (third term in equation (4.7)) of the volume fraction that takes into account the relative velocity between the two phases. The interface compression factor was set to its default value of 1 for all multiphase simulations in this thesis.

Due to the interface capturing method implemented in interFoam, the exact location and shape of the interface is unknown. Therefore the surface tension force cannot be directly calculated from the curvature but is instead modelled using the continuum surface force model of Brackbill et al. [6]. This model adds surface tension as a volume force $F_\sigma'$ to the momentum equation within the interfacial region using local gradients of the normal vectors of the interface to determine the mean curvature $H'$ [4, 60]:

$$F_\sigma' = \frac{2\sigma H' \rho}{\rho_L + \rho_G} \frac{\partial \beta}{\partial x_i'} \approx \sigma H' \frac{\partial \beta}{\partial x_i'} \quad . \tag{4.10}$$

The discretized partial differential equations for conservation of mass and momentum are solved using the PIMPLE (Pressure Implicit Method for Pressure Linked Equations) scheme, which is a combination of the steady-state SIMPLE and the transient PISO [25] (Pressure Implicit with Splitting of Operators) schemes. PIMPLE combines the advantages of under relaxation within SIMPLE with the transient solution capability of PISO to achieve faster convergence within a time step and allow for larger time steps. OpenFOAM allows adjustable time steps during the computation and the size if each step is calculated by the velocity-based and interface-based Courant numbers[1] which were limited to 0.2 and 0.1 respectively.

# 4.1   Mesh Generation

The computational domain of the CFD simulations is modelled to capture the geometry of the experiment setup in and around the test channel. As displayed in figures

---

[1]The Courant number is defined for one cell as $\mathrm{Co} = \frac{\delta t' |v_i'|}{\delta x_i'}$.

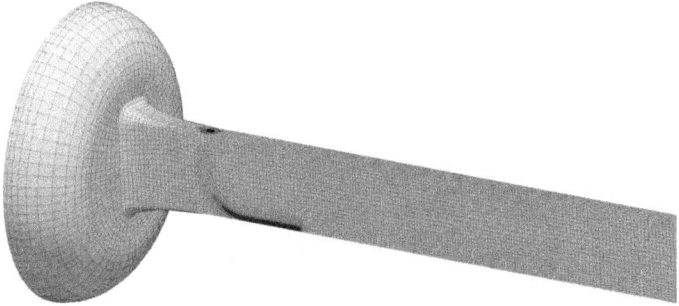

Figure 4.1: Structured mesh used for determining for the Reynolds number dependent pressure boundary condition using simpleFoam. Mesh analysis simulations were performed without insertion of the bubble injection cannula with snappyHexMesh.

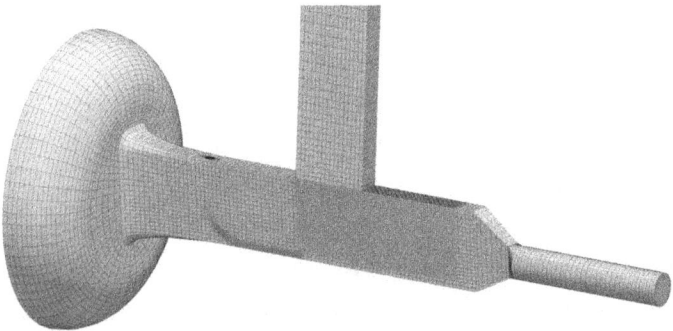

Figure 4.2: Example of structured mesh used for CFD simulations with interFoam. The length of the top rectangular segment depends on $l'$. Additional mesh refinement is applied in the vicinity of the free surface.

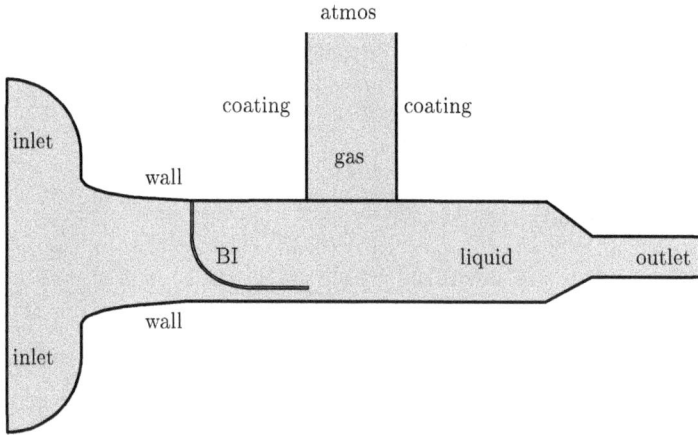

Figure 4.3: Boundaries and regions of computational domain for CFD simulations. The bubble injection cannula (BI) is included in the sketch and is assigned the same numerical properties as the channel walls.

3.5 and 3.6, flow enters into the test channel after passing through the FPC and an entrance nozzle. Within the entrance nozzle, a bubble injection cannula penetrates the flow path. Downstream of the test channel, the triangular duct converges to a circular pipe. Examples of the numerical grids that were used for simulating the flow problem are displayed in figures 4.1 and 4.2. The inlet of the computational domain is located in the FPC and corresponds to the effective flow area that is available through the porous screen (compare section 3.1). The outlet of the computational domain is located in the circular duct downstream of the TC. Multiple meshes for two-phase flow simulations were generated to vary the length of the test channel. The ambient gas region is modeled immediately above the free surface for two-phase flow simulations.

The computational grids were created by initially building a hexahedral structured mesh with the mesh generation tool gmsh [26] and subsequently converted to an OpenFOAM compatible format using the utility gmshToFoam. After local mesh refinement in the vicinity of the gas/liquid interface, the bubble injection cannula is inserted into the mesh using the OpenFOAM utility snappyHexMesh. Essentially,

snappyHexMesh is used to delete cells from the mesh based on a location map that
is defined by a three-dimensional model of the cannula. Before deleting the cells
from the mesh, the cell refinement level is increased in the vicinity of the cannula.
After cell deletion, the cell edges are fitted to the original shape of the cannula's
model to prevent sharp edges and skewed cells.

For single-phase flow simulations a mesh quality study was performed to determine the optimal mesh resolution for computing the pressure loss along the flow
path. The mesh quality study was performed without including the bubble injection
cannula. Having chosen the optimal mesh resolution based on solution accuracy and
computational time, the mesh was modified to include the cannula. The boundaries
of the meshes are outlined in figure 4.3. For single-phase simulations the duct was
closed (no gas region) and the duct did not converge to a circular cross-section at
the outlet.

## 4.2  Simulations with SimpleFoam

The boundary conditions of the single-phase flow simulations are outlined in table
4.1. Single-phase flow simulations were performed to determine the quality of the
generated mesh (compare figure 4.1) and to calculate the pressure loss as a function
of the Reynolds number. The pressure loss function is required for the pressure
boundary condition of the one-dimensional model and is discussed in section 2.2.7.

Table 4.1: Boundary conditions for the single-phase simulations.

|        | inlet    | outlet                          | wall                            |
|--------|----------|---------------------------------|---------------------------------|
| $v_i'$ | constant | constant                        | 0                               |
| $p'$   | 0        | $\dfrac{\partial p'}{\partial x_i'} = 0$ | $\dfrac{\partial p'}{\partial x_i'} = 0$ |

### 4.2.1 Mesh Quality Analysis

As mentioned above, the mesh quality study was performed for single-phase flow through the closed test channel using the OpenFOAM solver simpleFoam for a variety of mesh resolutions. Due to the absence of a phase interface, no mesh refinement was applied inside the computational grid. Also, the bubble injection cannula was not included in the mesh. The mesh resolution was varied by increasing the number of cells in across the width (along the $y$-axis) of the TC. The mesh resolution in the other spatial coordinates depended on the cell length along the $y$-axis. Six mesh resolutions were examined and the average pressure in the cross-section of the TC's inlet ($x = 0$) was compared. The total number of cells ranged from 31 000 to 234 000 and the effect of the mesh resolution on the pressure loss is displayed in figure 4.4. The results apparently converge toward an accurate solution with increasing number of cells. The sufficiently adequate mesh for the pressure loss computations was chosen to be the mesh with $N_{cells} \approx 10^5$. The decision was based on the relative accuracy and the computation time.

## 4.3 Simulations with InterFoam

### 4.3.1 Boundary Conditions

The boundary conditions applied in the two-phase simulations are summarized in table 4.2. The pressure in the gas phase is constant at ambient pressure $p'_a$. The gas region is enclosed by walls and has an atmospheric boundary condition imposed on its lid. The pressure at the inlet of the computational domain is defined by the capillary pressure imposed by the meniscus in the CT, which is not included in the mesh, but yields:

$$p'_{FPC} = p'_a - \frac{2\sigma}{R'_{CT}} = p'_a - 1.0773\,\text{Pa} \quad , \tag{4.11}$$

which is also the pressure within the liquid when it is not flowing.

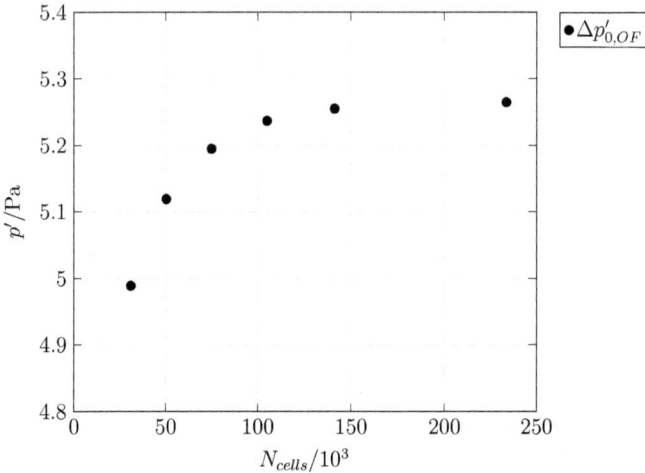

Figure 4.4: Computed pressure loss at varying levels of mesh resolution for single-phase flow. The results apparently converge toward an accurate solution with increasing number of cells.

The velocity of the flowing liquid is defined by a constant (or ramped) flow rate boundary condition at the outlet to simulate the pump in the experiment setup. A no-slip boundary condition is imposed on the walls of the channel (the BI cannula has identical boundary conditions as the walls). All walls have a zero contact angle boundary condition except those enclosing the gas region, where the contact angle is defined as 180 degrees to avoid unphysical behaviour at the pinning edges of the TC. Pressure, velocity, and volume fraction conditions in the liquid and gas domains are defined before initiating the computation using the OpenFOAM utility setFields. The physical properties of the numerical fluids are defined as the properties of HFE7500 and nitrogen at 303.15 K, which are displayed in table 3.2 for the liquid. The properties of gaseous nitrogen at 303.15 K are computed according to Lemmon et al. [37] and are defined as constant values $\rho_G = 1.11 \, \mathrm{kg \, m^{-3}}$ and $\nu_G = 1.62 \times 10^{-5} \, \mathrm{m^2 \, s^{-1}}$ in the simulations.

Table 4.2: Boundary conditions for the two-phase simulations. Open-
FOAM boundary conditions are abbreviated as follows: zG=zeroGradient;
fFP=fixedFluxPressure; iO=inletOutlet.

|  | inlet | outlet | wall | coating | atmos |
|---|---|---|---|---|---|
| $v_i'$ | zG | $\dfrac{Q'}{A_0'}$ | 0 | 0 | zG |
| $p'$ | $p_a' - p_{CT}$ | zG | fFP | fFP | $p_a'$ |
| $\beta$ | $\beta = 1$ | zG | $\gamma = 0°$ | $\gamma = 180°$ | iO |

## 4.3.2 Simulation Procedure

Two-phase flow simulations were performed to determine $Q_{crit,3D}'$ for single-phase
flow through the TC with a gas/liquid interface in the TC. Simulations were performed in two stages for each examined channel length and essentially differed only
in the velocity boundary condition. In the first stage a ramp boundary condition
was imposed on the flow rate at the outlet. A total of 50 seconds was simulated, in
which the flow rate boundary condition at the outlet $(Q_{out}')$ was increased steadily
from Zero to $5 \times 10^{-6}\,\mathrm{m^3\,s^{-1}}$. A first approximation of $Q_{crit,3D}'$ was then determined
by computing the flow rate difference between the outlet and inlet[2] of the test chan-
nel $(\Delta Q_{3D}' = Q_{out}' - Q_{in}')$. When choking occurs, the inlet is no longer able to supply
the flow rate demand at the outlet of the channel due to the flow rate limitation
within the TC. Therefore, a difference between the two flow rates marks the onset
of choking. As can be seen in figure 4.5, $\Delta Q_{3D}' \approx 0$ until around $t' = 35\,\mathrm{s}$ and
$Q_{out}' = 3.5 \times 10^{-6}\,\mathrm{m^3\,s^{-1}}$ for a channel length of $l' = 25 \times 10^{-3}\,\mathrm{m}$.

Following the experiment procedure, the next stage involves performing simula-
tions at constant flow rate demands in steps of $0.1 \times 10^{-6}\,\mathrm{m^3\,s^{-1}}$ around the approximate $Q_{crit,3D}'$. Each case simulates 10 seconds of real time and for each simulation
the stability of the interface in the TC is examined to determine whether the flow

---

[2]$Q_{in}'$ is computed according to the zero gradient boundary condition at the inlet.

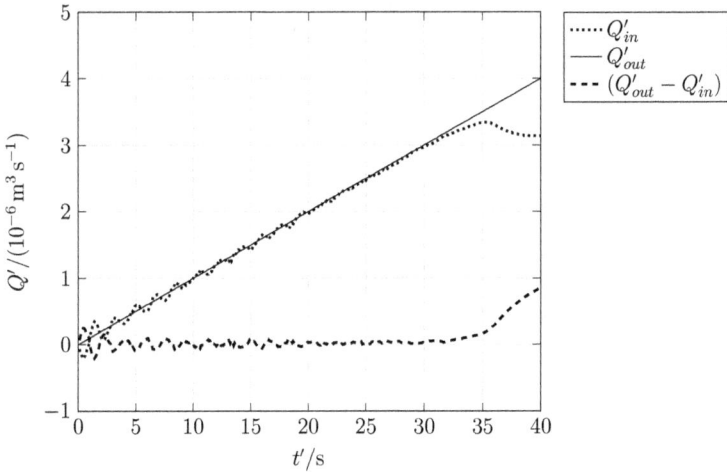

Figure 4.5: Comparison of inlet flow rate against outlet flow rate boundary condition over time in two-phase flow simulations for $l' = 25 \times 10^{-3}$ m. The approximate onset time of choking can be interpreted from the difference of the two flow rates $Q'_{out}$ and $Q'_{in}$.

in the TC is choked. Examples of subcritical and supercritical cases are shown in figure 4.6.

### 4.3.3   Representative Results

Critical flow rates were determined from CFD simulations for $5 \times 10^{-3}$ m $\leq l' \leq 48 \times 10^{-3}$ m and are displayed in table 4.3. Simulations were performed at increasingly higher flow rates until choking was observed in the simulation results. Flow rate intervals between simulations are $SI0.05e - 6m^3.s^{-1}$. The obtained results are compared with experiment results in section 6.2.

In addition, free surface contours were evaluated for both critical and a variety of subcritical flow rates. The contours of the free surface were determined by averaging

(a) (b)

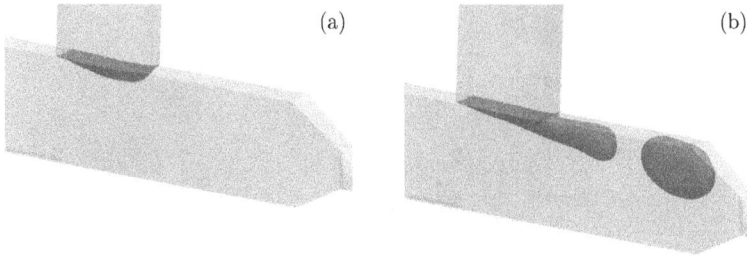

Figure 4.6: Comparison of stable interface (a) and choked flow with bubble ingestion (b) in numerical simulations. The phase interface is colored dark grey. The highest steady flow rate at which the interface remains stable is defined as $Q'_{crit,3D}$.

the isoline of $\beta = 0.5$ in the test channel over 10 s or 20 s of simulated time. As displayed representatively in figure 4.7, flow separation was observed immediately downstream of $x'(k'_{min}) = x'^*$ in the CFD simulations[3]. The flow separation results in a recirculation zone downstream of the free surface. The eddies appear three-dimensional in nature, qualitatively similar to vortices behind a spherical obstacle in low Reynolds number flow. No quantitative investigation of the flow separation and eddies was performed. It is however assumed that the flow separation contributes to the pressure differential along the length of the channel. Flow separation is not considered in the presented one-dimensional model and is therefore not taken into account in the prediction of the critical flow rate in open capillary channel flow.

For the one-dimensional model the viscous losses in the channel are estimated based on the assumption that the velocity profile is fully developed at the inlet of the test channel. According to the performed CFD simulations, this is a false assumption. Figures 4.8, 4.9, and 4.10 clearly show that the entrance length upstream of the test channel's inlet is too short for the velocity profile to develop fully. Especially figure 4.8 demonstrates this observation via the scaled stream-wise velocity along the $y$-axis at $x_0$. For decreasing Re, the velocity profile approaches the expected profile

---

[3]$x'^*$ is the location along $x'$ in the test channel with the smallest cross-sectional area in the $y'z'$-plane

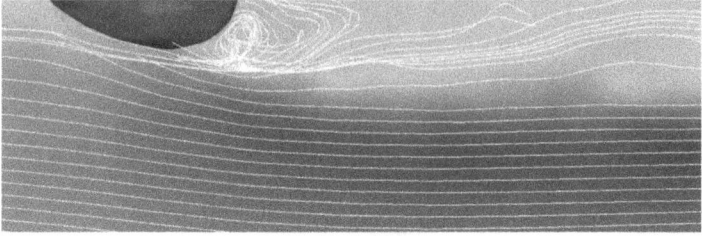

Figure 4.7: Evident flow separation at the free surface in CFD simulations. In this case, $l' = 10 \times 10^{-3}$ m and $Q' = 3.85 \times 10^{-6}$ m$^3$ s$^{-1}$ (subcritical flow). The interface is coloured grey, the velocity magnitude is coloured from red to blue (high to low), flow is from left to right. Streamlines are displayed in white. The side view is cropped. A recirculation zone is observed immediately downstream of $x'^*$. The eddy is three-dimensional in nature, which causes the illusion of crossing streamlines from this particular perspective.

that has a scaled maximum velocity of $v'_{max}/\bar{v}' = 2.432$ according to the work of Shah and London [51]. Therefore, an additional viscous term for pressure loss in the test channel must be missing in the one-dimensional model but is accounted for in CFD results, which may lead to diverging results especially in the lower range of Re observed here.

Also, as already debated by Klatte [33], the sudden change of the wall boundary condition from no-slip to slip at the inlet of the test channel may cause additional pressure loss due to the re-development of the flow profile because the fully developed profiles of a closed channel and an open channel differ from each other significantly.

Table 4.3: Critical flow rates determined with 3D CFD simulations in the channel length range observed in the experiments on the ISS and Oh = $2.485 \times 10^{-3}$.

| $l'/(10^{-3}$ m$)$ | 5 | 10 | 15 | 20 | 25 | 30 | 35 | 40 | 45 | 48 |
|---|---|---|---|---|---|---|---|---|---|---|
| $Q'_{crit,3D}/(10^{-6}$ m$^3$ s$^{-1})$ | 5.2 | 4.1 | 3.7 | 3.4 | 3.2 | 3.1 | 3.1 | 3.0 | 3.0 | 2.9 |

The CFD simulations performed in this work also unveil an additional deviation from the assumption made in the one-dimensional model that is caused by a disturbance of the velocity field as liquid passes around the bubble injector. As shown in figure 3.6, the cannula of the BI enters the closed channel upstream of the inlet at the base of the triangular cross-section and obstructs the entire height of the flow path as it makes its way to the test channel's inlet. The influence of the flow disturbance on the velocity profile at the inlet of the test channel is displayed in figures 4.9 and 4.8. For Re = 50, this disturbance may be neglected, but as Re increases, the disturbance is more pronounced and has a larger effect on the friction factor within the test channel. Given that Re > 200 for the entirety of the flow rates discussed in this thesis, the observations made from CFD simulations concerning the velocity profile should also be taken into consideration when determining the accuracy of the one-dimensional model.

In summary, although the pressure loss is well modelled due to the fit described in section 2.2.7, the CFD simulations reveal that the friction factor that is used to model viscous pressure loss in the one-dimensional model may be significantly lower than required to approximate the viscous pressure loss in CFD simulations and experiments. In chapter 6 the results of the three-dimensional simulations are compared to experiment results and those of the one-dimensional model.

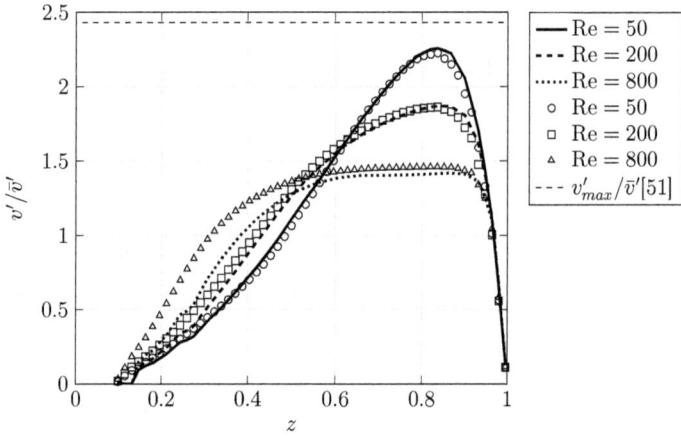

Figure 4.8: Dimensionless velocity profiles at $x_0$ for Re $= \{50; 200; 800\}$ with (lines) and without (marks) BI present in the control volume. All velocities are determined at $y = 0$. The horizontal dashed line represents the expected $v'_{max}/\bar{v}'$ according to Shah and London [51].

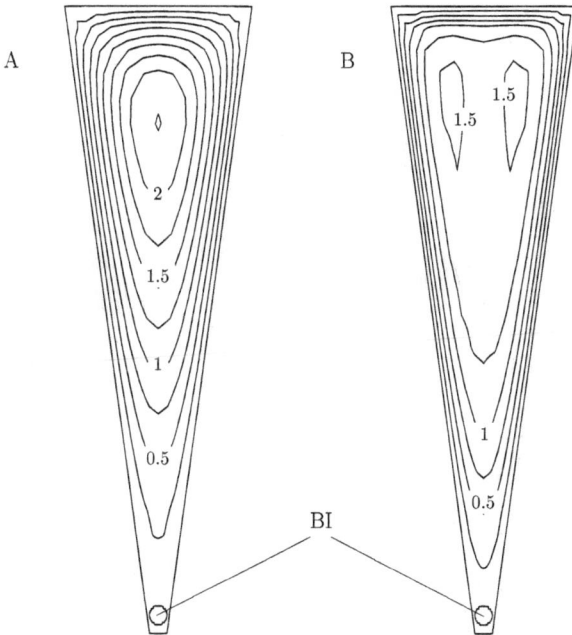

Figure 4.9: Dimensionless velocity profiles at the inlet of the test channel for $Re = 50$ (A) and $Re = 800$ (B). Velocity is scaled here by the respective average velocity in the test channel $\bar{v}' = Q'/A_0'$. Isolines describe the velocity profile and are plotted in increments of $0.5\,\bar{v}'$. The velocity profile appears almost fully developed in (A) whereas the profile in (B) displays irregularities that stem from flow disturbances upstream. The outlet of the bubble injector (BI) is located at the inlet of the test channel and is visualized here.

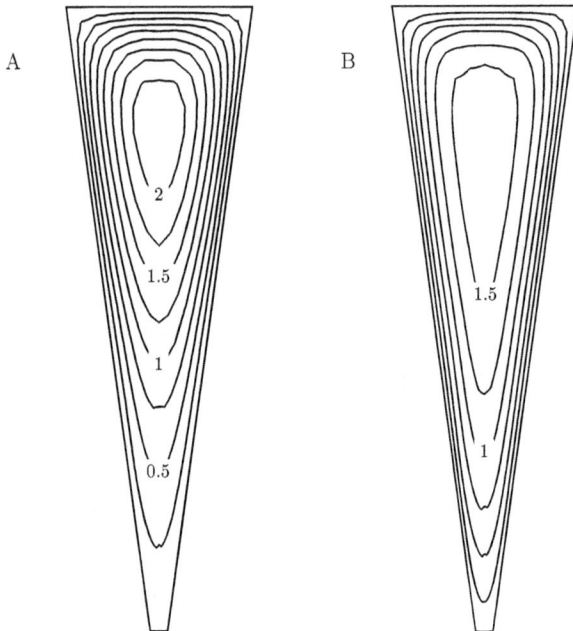

Figure 4.10: Dimensionless velocity profiles at the inlet of the test channel for Re = 50 (A) and Re = 800 (B) without a BI. Velocity is scaled here by the respective average velocity in the test channel $\bar{v}' = Q'/A_0'$. Isolines describe the velocity profile and are plotted in increments of 0.5 $\bar{v}'$. The velocity profiles display varying degrees of profile development but no disturbances.

# Chapter 5

# One-Dimensional Numerics

The governing equations of the one-dimensional model presented in chapter 2 are solved numerically using an in-house developed software tool called ccflow. The tool is written in FORTRAN and C and is capable of producing numerical predictions for critical flow rates in steady or transient flow through channels with a rectangular cross-section and steady flow through channels with a triangular cross-section. In the process, the flow variables local mean velocity, cross-sectional area, curvature and height of the free surface, and pressure are computed and recorded for comparison with data from alternative sources.

The governing equations are integrated into the programme in dimensionless form (see equations (2.70), (2.71), (2.72), and (2.73)) and are approximated using the method of finite differences. The user may choose between a first order up-wind discretization scheme or the more accurate but less robust central differences discretization scheme.

The control volume that is modelled is restricted to $0 \leq x \leq l$, which means that the interface is bounded to the confines of the test channel. In most cases, this is an acceptable assumption; however, as Klatte [33] shows, cases may arise where the interface extends beyond $l$. This is mostly the case for supercritical flow $(Q > Q_{crit})$, a parameter range that lies outside the computation capabilities of ccFlow. The channel length is discretized into a user-defined number of grid points that is generally chosen to be between 200 and 3500.

For steady flow problems, such as the one regarded in this thesis, the set of coupled, non-linear partial differential equations is solved with a damped Newton method for the variables $v, A, k, h$ for constant $Q$ [28]. The specific algorithm used in the programme for steady flow is NLEQ (Non -Linear EQuations), a non-commercial code for the solution of systems of highly nonlinear equations that is copyrighted by Konrad Zuse Zentrum fuer Informationstechnik Berlin [20, 41]. The critical flow rate is defined as the highest flow rate for which a converged solution for balanced steady flow can be found. A simple homotopy method is applied to determine $Q_{crit}$ within an interval $Q_1 \leq Q_{crit} \leq Q_2$, where $Q_1$ is a converged solution and $Q_2$ is a flow rate for which the algorithm was not able to find a converged solution. The interval is successively reduced until $Q_2 - Q_1 < 10^{-6}$, at which point $Q_{crit}$ is assumed to be sufficiently accurate and the solution is presented as $Q_{crit} = Q_1$.

ccFlow presents users with a graphical user interface to modify model parameters and adjust numerical settings. Model parameters may be entered either directly in non-dimensional form or in dimensional form and are subsequently scaled according to the method presented in section 2.4. Previous studies with ccFlow involving comparisons with data from experiments have shown that it is sufficiently accurate for the presented flow problem in rectangular channels [9, 28, 30] and that the accuracy is highest in flows where viscous forces are more dominant than inertia. The limited experimental data available for wedge-shaped channels has similarly shown that ccFlow is able to accurately predict the critical flow rate in this capillary channel geometry [33].

As yet, ccFlow is restricted to single-phase flow problems. For this thesis, the source code was modified to allow implementation of a homogenous two-phase flow model for dispersed bubbly flow (see section 2.3). The modifications accounted for bubble injection within the parameter range observed during the experiments that were performed on the ISS. Representative results of the single-phase one-

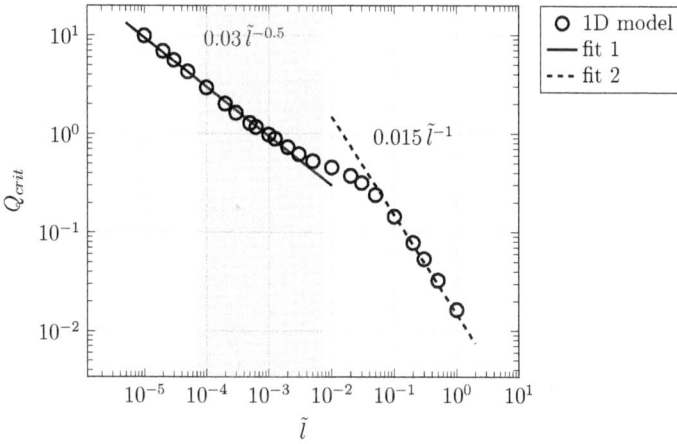

Figure 5.1: Numerical results for dimensionless critical flow rate $Q_{crit}$ over dimensionless channel length $\tilde{l}$ in a double logarithmic plot. The applied scaling reveals two distinct regimes that can be readily fitted with simple power functions. The proposed fit functions are plotted and labeled. The range of $\tilde{l}$ observed in the experiments is coloured grey.

dimensional model are presented in this section. A comparison of the numerical and experimental results, including numerical results for two-phase flow, is located in chapter 6.

# 5.1   Single-Phase Flow

Critical flow rates are determined for $Oh = 2.485 \times 10^{-3}$ and $\Lambda = 7.207$. The critical flow rate displays a strong dependency on the channel length which is apparent in dimensionless form in figure 5.1 for $10^{-5} < \tilde{l} < 1$. Upon closer observation the dependency of $Q_{crit}$ on $\tilde{l}$ displays a simple power law relationship for this particular set of model parameters. For inertia dominated flows, $Q_{crit}$ may be fitted to a $Q_{crit} \sim \tilde{l}^{-0.5}$ curve. For viscous dominated flow, the power law relationship may

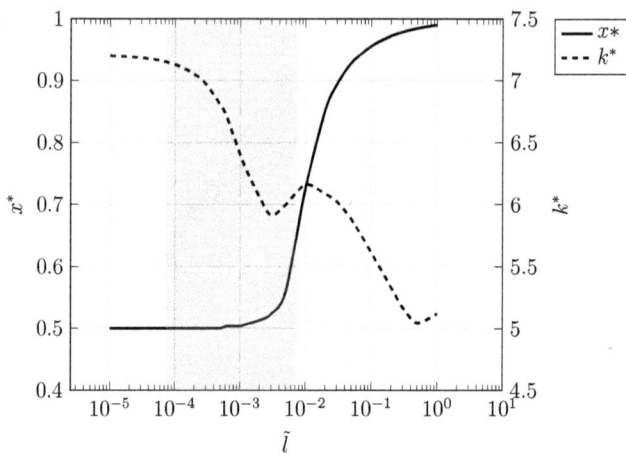

Figure 5.2: Numerical results for $k^*$ and $x^*$ at $Q_{crit}$ over $\tilde{l}$. The dependency on the dimensionless channel length implies a difference in the dominant pressure terms. The range of $\tilde{l}$ observed in the experiments is coloured grey.

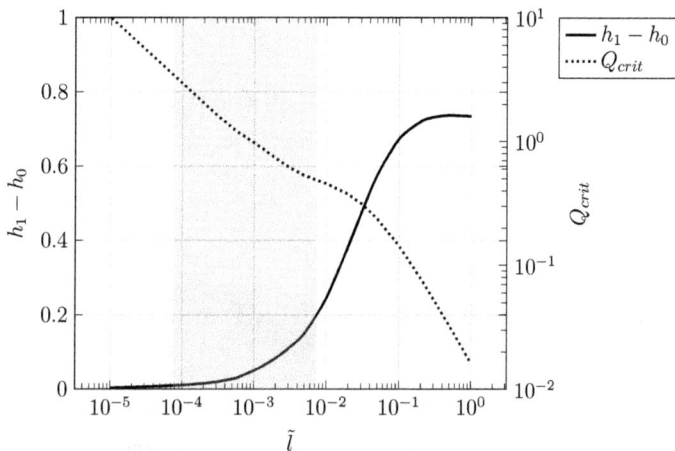

Figure 5.3: Numerical results for the dimensionless curvature difference $h_1 - h_0$ between outlet and inlet of the channel over $\tilde{l}$. The difference equals the modelled irreversible pressure loss. Numerical results for the dimensionless critical flow rate are plotted for comparison. The range of $\tilde{l}$ observed in the experiments is coloured grey.

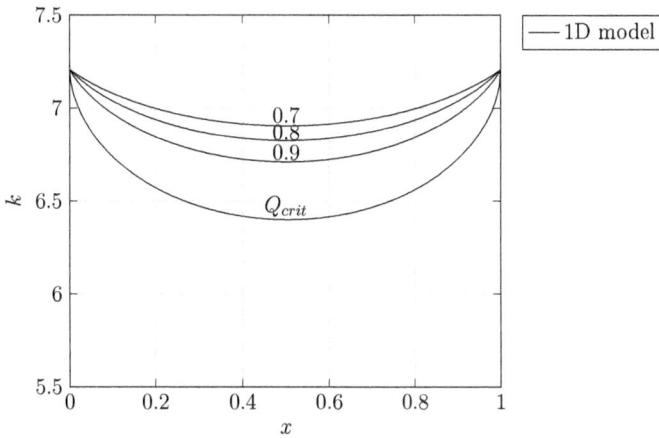

Figure 5.4: Dimensionless contour plots for Oh $= 0.002485$, $\Lambda = 7.2066$, and $\tilde{l} = 10^{-3}$ at flow rates $0.7\,Q_{crit} \leq Q \leq Q_{crit}$ and labelled accordingly.

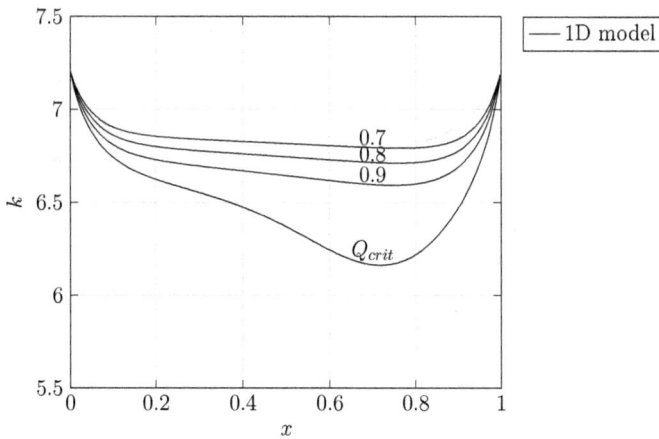

Figure 5.5: Dimensionless contour plots for Oh $= 0.002485$, $\Lambda = 7.2066$, and $\tilde{l} = 10^{-2}$ at flow rates $0.7\,Q_{crit} \leq Q \leq Q_{crit}$ and labelled accordingly.

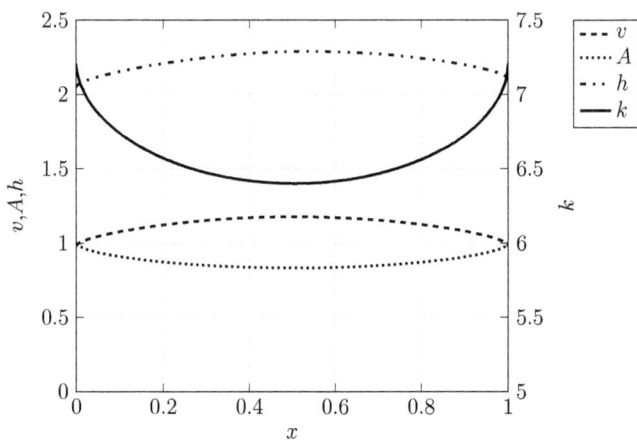

Figure 5.6: Numerical results for $v$, $A$, $h$, $k$ at Oh $= 0.002485$, $\Lambda = 7.2066$, and $\tilde{l} = 10^{-3}$.

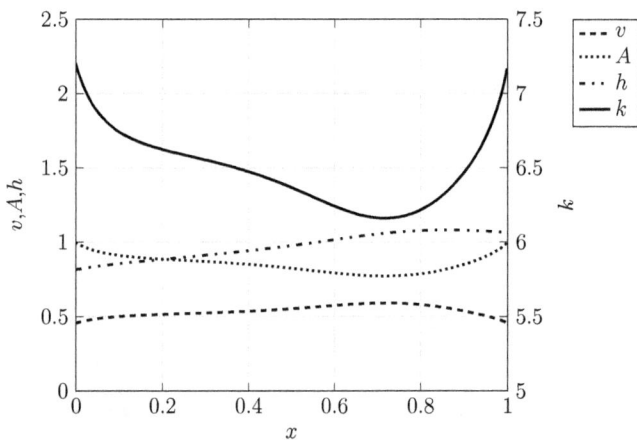

Figure 5.7: Numerical results for $v$, $A$, $h$, $k$ at Oh $= 0.002485$, $\Lambda = 7.2066$, and $\tilde{l} = 10^{-2}$.

be fitted to a $Q_{crit} \sim \tilde{l}^{-1}$ curve. A transition interval is apparent in the range of $10^{-3} < \tilde{l} < 10^{-1}$. Interestingly, the difference in the exponents displays a factor of 2 between the two cases, but this may be coincidental and a result of the particular model parameters Oh and $\Lambda$ which were chosen to be identical to the parameters that were observed in the experiments.

In the presented one-dimensional model, the momentum equation (equation (2.70)) balances the capillary pressure difference against the sum of the convective pressure gradient and viscous pressure loss and is valid for $Q \leq Q_{crit}$. In fully developed flow, as is assumed here, the convective pressure at the inlet and outlet of the channel are equal in accordance with streamtube theory. This means that a pressure difference between $x = 0$ and $x = 1$ must be equal to the irreversible pressure loss that is included in the model and may be interpreted as the difference of the curvatures $h(x = 1)$ and $h(x = 0)$ [30]. This reasoning may be exploited to determine whether convective or viscous forces are dominant for the pressure development along the channel. Figure 5.3 displays how the dimensionless curvature difference $h(x = 1) - h(x = 0)$ depends on the dimensionless channel length $\tilde{l}$. It can be seen that for $\tilde{l} < 10^{-3}$ ($l' = 6.6 \times 10^{-3}\,\text{m}$) the viscous pressure loss plays a relatively insignificant role. For $\tilde{l} > 10^{-3}$, the irreversible pressure loss rises sharply and gains importance. The interval $10^{-3} < \tilde{l} < 10^{-1}$ marks a transition zone in which both inertia and irreversible pressure loss are of importance and must be regarded. For channel lengths $\tilde{l} > 10^{-1}$ ($l' = 0.66\,\text{m}$) the viscous pressure loss is the dominant term in the balance equation. Based on these observations, the respective terms may be neglected in the model in the appropriate parameter ranges, but are included at all times in the present thesis.

Additional data from the model may be derived in the form of contour data describing the height $k(x)$ of the interface in the symmetry plane. Significant differences are observed here also between the two distinct physical regimes of convective and viscous dominated flow in the aforementioned intervals of $\tilde{l}$. In figure 5.4, contour data for $\tilde{l} = 10^{-3}$ is presented representatively for inertia dominated flows at various factors of $Q_{crit}$. The contour appears symmetrical with a lowest point

$\min(k(x)) = k^*$ at $x = 0.5$ (or in dimensional form $x'^* = 0.5l'$). As $\tilde{l}$ increases, the contour appears less symmetrical and $x^*$ shifts farther towards the outlet of the channel. This is displayed clearly in figure 5.2, where $k^*$ and $x^*$ are plotted for a wide range of $\tilde{l}$ at $Q_{crit}$. The dimensionless variables $k^*$ and $x^*$ both display distinct differences between the aforementioned physical regimes.

Also of note is the presence of an inflection point in the contour of the interface for cases with $\tilde{l} \geq 1 \times 10^{-1}$ as displayed in figure 5.5 which shows the interface contour at various factors of $Q_{crit}$ representatively for viscous dominated flows. For all cases, an increase of sensitivity of the interface is displayed as $Q$ approaches $Q_{crit}$ which is shown by the rate at which $k^*$ changes for constant steps of constant $Q$ (compare figures 5.4 and 5.5). Further variables that display dependencies on $\tilde{l}$ are displayed in figures 5.6 and 5.7, where $A$, $k$, $v$, and $h$ are plotted over $x$ for the cases $\tilde{l} = 10^{-3}$ and $\tilde{l} = 10^{-2}$ respectively. Comparing the development of $h$ over $x$ between the two cases, the relative dominance of the convective and viscous terms once again become apparent. For $\tilde{l} = 0.001$ the lost pressure is almost completely restored in accordance with streamtube theory. For $\tilde{l} = 0.01$, however, a clear difference between $h(x = 0)$ and $h(x = 1)$ is observed and can be attributed to the irreversible viscous pressure loss included in the model.

## 5.2   Two-Phase Flow

Critical flow rates for two-phase flow are calculated based on the one-dimensional model outlined in sections 2.2 and 2.3. Results are computed for channel lengths $l' = \{10, 15, 20, 25, 30\} \times 10^{-3}$ m and averaged gas flow rates $Q'_G = \{0.1, 0.2, 0.3, 0.4\} \times 10^{-6}$ m$^3$ s$^{-1}$. The calculated critical flow rates are displayed in figure 5.8. The curves show that gas injection generally leads to an increase or an enhancement of the critical flow rate and that this enhancement depends on $Q'_G$. The quantitative enhancement due to gas injection is defined as the difference between the two-phase critical flow rate and the single-phase critical flow rate ($\Delta Q'_{crit} = Q'_{crit,2P} - Q'_{crit,1P}$), which is plotted over the injected gas flow rate in figure 5.9, where the dependency of $\Delta Q'_{crit}$ on $Q'_G$ is apparent. A possible dependency of $\Delta Q'_{crit}$ on the channel length $l'$ does not appear to be nearly as significant as $Q'_G$. Enhancement of the critical flow rate is consistently observed to be higher than the injected gas flow rate in the model's predictions, therefore it is assumed that gas injection within the observed parameters has a stabilizing effect on the free surface in the test channel. A comparison of the predicted two-phase critical flow rates with experiment results is found in chapter 6.

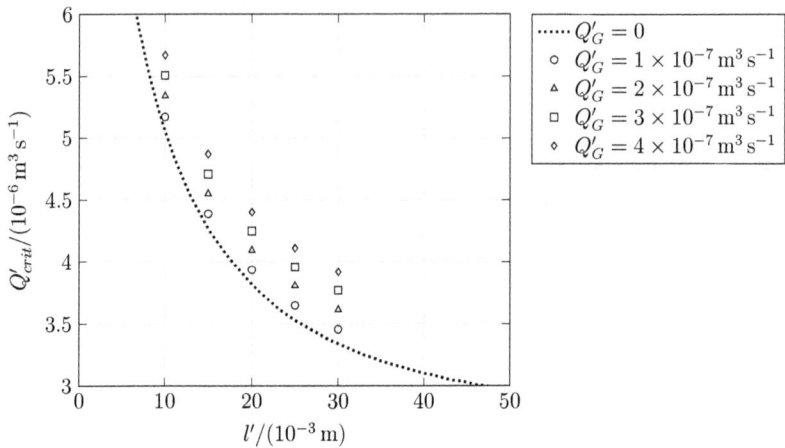

Figure 5.8: Critical flow rates in two-phase flow for various gas injection flow rates.

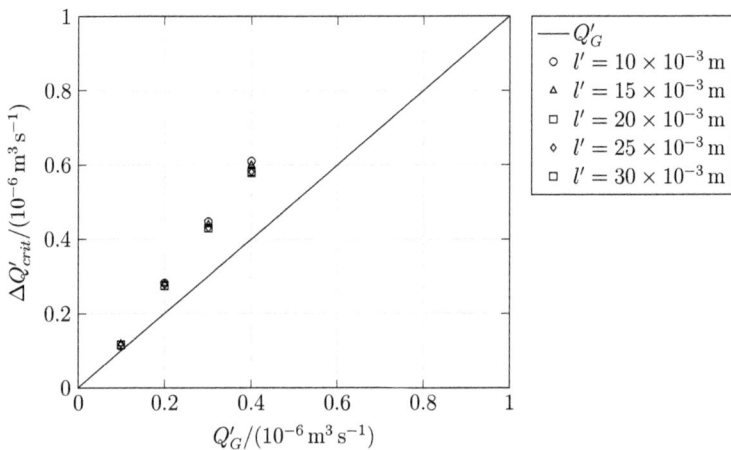

Figure 5.9: Critical flow rate enhancement versus gas flow rate for one-dimensional numerical simulations of bubbly two-phase flow.

# Chapter 6

# Comparative Results and Discussion

Hardware installation onboard the ISS took place on December 27, 2010 for EU1 and operations lasted until March 17, 2011. The first installation of EU2 was completed on September 19, 2011 and operations lasted until October 17, 2011. Additional operations windows were available in 2012 and 2013 and were used to perform more experiments with the EU2 setup. Installation onboard the ISS was performed excellently by various NASA astronauts. The installation procedure required less than 3 hours of crew time for each EU. After hardware installation, no further crew involvement was required for nominal operations until the end of operations. Then the experiment was removed from the MSG and placed in stowage onboard the ISS. Prior to removal from the MSG, the setup was returned to its pre-installation status as well as possible. This involved removing gas from the PSC, filling the CT with liquid and closing all valves. The test channel was not drained for stowage.

The experiment hardware was controlled from two ground stations, one located at ZARM in Bremen, Germany, the other at Portland State University in Portland, Oregon, USA. This enabled the research team to gather data around the clock with the research teams on either side of the Atlantic taking turns to command the experiment. The data connection to the experiment was reasonably direct with only a few seconds passing from initiating a command from the ground station and observing the response in the telemetry data and the video feed. The experiment hardware, especially the fluid loop, proved resilient against oscillations and accelerations that were caused by manoeuvres of the ISS, but no science data was collected when such disturbances of microgravity occurred. The combination of pump, sensors, plungers, and valves were also used to maintain the fluid loop when necessary. They were also used to adjust the system pressure, the fill level in the CT, and the size of the bubble in the PSC.

Commanding of the experiment hardware and reception of its telemetry data were limited to periods of acquisition of signal (AOS), when a data connection to the ISS was available. During loss of signal (LOS) no data was received and the ability to command the experiment was suspended by NASA ground control. Therefore no experiments took place during these periods. Prior to scheduled LOS phases the experiment was set to a 'safe' mode, in which test channel is closed and a low flow rate is established. The duration of AOS periods was typically around one or two hours with short LOS intervals of less than an hour between them, but these durations differed based on current operations on the ISS and satellite coverage. Furthermore, experiments were suspended when docking or re-boost procedures were performed onboard the ISS.

## 6.1   Behaviour of the Free Surface

The behaviour of the free surface in the experiments at steady flow rates can be divided into two categories, stable (for $Q' \leq Q'_{crit}$) and choked (for $Q' > Q'_{crit}$). Examples of both categories are given in figures 6.1 and 6.2. The free surface is visible as a black area on the open sides of each channel. The height of the free surface, $k'$, is an indication of the pressure difference between the flowing liquid in the test channel and the ambient gas. In figure 6.1, the flow rate has not exceeded $Q'_{crit}$. Minor oscillations can be observed on the free surface which are believed to emanate from vibrations caused by the gear pump that generates the flow. Besides

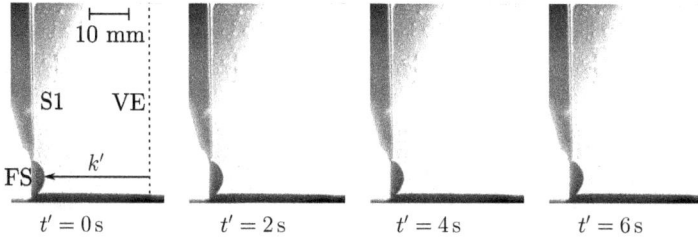

Figure 6.1: Selected images from experiment WE1125 showing stable flow from bottom to top in the wedge-shaped channel at $Q' = 3.87 \times 10^{-6}\,\mathrm{m^3\,s^{-1}}$ with $l' = 10 \times 10^{-3}\,\mathrm{m}$. The vertex (VE) of the wedge is indicated by the dashed line. The position of the free surface within the channel is indicated as $k'$. The stable free surface is visible as the black area bending into the channel at the lower left side of the image.

these oscillations the free surface is stable throughout the entire duration of the experiment in each of the geometries.

In figure 6.2 an example is given for choked flow in the test channel. In each of the experiments the flow rate has exceeded $Q'_{crit}$. In these cases the maximum mean curvature of the free surface is no longer sufficient to balance the pressure difference between the liquid in the channel and the ambient gas pressure. The free surface continues to bend further into the channel until a gas bubble is ingested into the liquid. Immediately after gas ingestion the free surface snaps back and the process repeats itself. In the groove and wedge-shaped channels ingestion is only possible on one side because only one free surface is present. However, in the parallel plate channel gas ingestion is possible on either side.

Images recorded by the HSHR camera display artifacts that appear as specks on the test channel and can be seen in figure 6.2. The source of these specks is unknown and they were observed before the initial filling procedure. The visibility of the artifacts depended on the lighting conditions and although the degradation of the image quality is a nuisance, the flow within the test channel was not affected. The number of specks and their location was impervious to flow conditions within

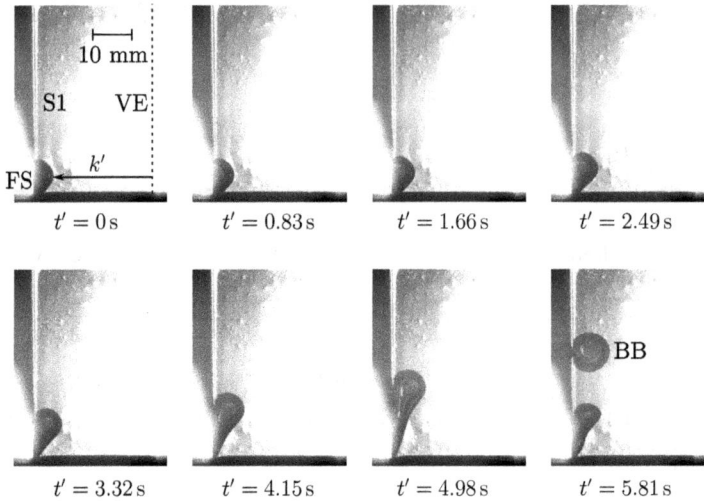

Figure 6.2: Selected images from experiment WE1433 showing choked flow from bottom to top in the wedge-shaped channel at $Q' = 3.93 \times 10^{-6} \, \mathrm{m^3 \, s^{-1}}$ with $l' = 10 \times 10^{-3} \, \mathrm{m}$. The vertex (VE) of the wedge is indicated by the dashed line. During the experiment the free surface proceeds to bend further into the channel. As soon as gas is ingested as a bubble (BB) the free surface begins to retract.

the TC. These observations lead the authors to the conclusion that the source of the specks is located outside the test channel and can therefore be neglected.

## 6.2   Single-Phase Critical Flow

Critical flow rates were determined experimentally for steady flow conditions in each channel geometry in the range of $10^{-3} \, \mathrm{m} \leq l' \leq 48 \times 10^{-3} \, \mathrm{m}$. The results have been previously presented in Canfield et al. [13] and are displayed here in figure 6.3. The error of the experiment results may be defined as the accuracy of the flow meter (see section 3.2) or less considering the repeatability of the determined values and the low standard deviation of the averaged critical flow rates (compare figure 6.4). For most channel lengths $Q'_{crit}$ was determined multiple times to ensure

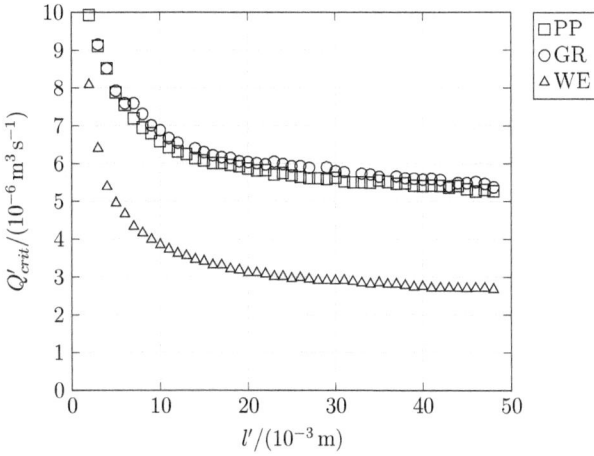

Figure 6.3: Experimentally determined critical flow rates for steady flow for the individual test channel shapes: parallel-plates (PP), groove (GR), and wedge (WE).

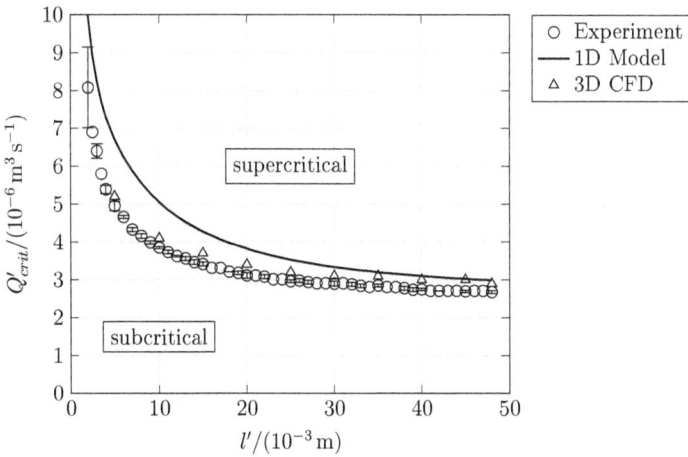

Figure 6.4: Comparison of experiment results for $Q'_{crit}$ with those of 1D and 3D simulations for steady single-phase flow in the wedge. Standard deviations are displayed as error bars for averaged experiment results.

repeatability of the results. The standard deviation between the observed $Q'_{crit}$ for a given channel length is generally lower than the accuracy of the flow meter. All three geometries display decreasing values for $Q'_{crit}$ with increasing $l'$. For short channels with $l' < 30 \times 10^{-3}$ m the decline of $Q'_{crit}$ appears hyperbolic. For longer channels with $l' \geq 30 \times 10^{-3}$ m the decline of $Q'_{crit}$ appears to be linear. As discussed in Rosendahl et al. [46] and in chapter 5, this could be explained by the transition from an inertia dominated flow regime towards a viscous dominated flow regime as the flow rate decreases.

The difference between the channel geometries is most significant in the wedge. The values of $Q'_{crit}$ are significantly lower in the wedge-shaped channel than in the parallel-plates or groove channels, which, in comparison, hardly differ from each other (on average a difference of approximately 3 %). The highest critical flow rates attained at a given channel length within the observed range are found in the groove geometry.

The experiment results obtained with the wedge geometry are compared with the numerical results for the values of the critical flow rate in figure 6.4 and in table 6.1. The CFD computations agree very well with the experiment results. The critical flow rates of the 3D CFD computations are slightly higher than the experiment results throughout the study and display an almost constant positive offset when compared to $Q'_{crit,exp}$ (on average about $0.25 \times 10^{-6}$ m$^3$ s$^{-1}$). A relative error between the experiment results $Q'_{crit,exp}$ and the numerical results $Q'_{crit,1D}$ and $Q'_{crit,3D}$ is defined here as:

$$\varepsilon_{xD} = \left| Q'_{crit,exp} - \frac{Q'_{crit,xD}}{Q'_{crit,exp}} \right| \quad , \tag{6.1}$$

where the index $xD$ may be substituted with $1D$ or $3D$ for the respective numerical results from one-dimensional and three-dimensional computations. The relative error of the 3D simulations is $\varepsilon_{3D} < 10$ % across the entire range of examined channel lengths.

The results of the numerical simulations with the 1D model also display satisfactory levels of agreement with the experiment results. In fact, for channels longer

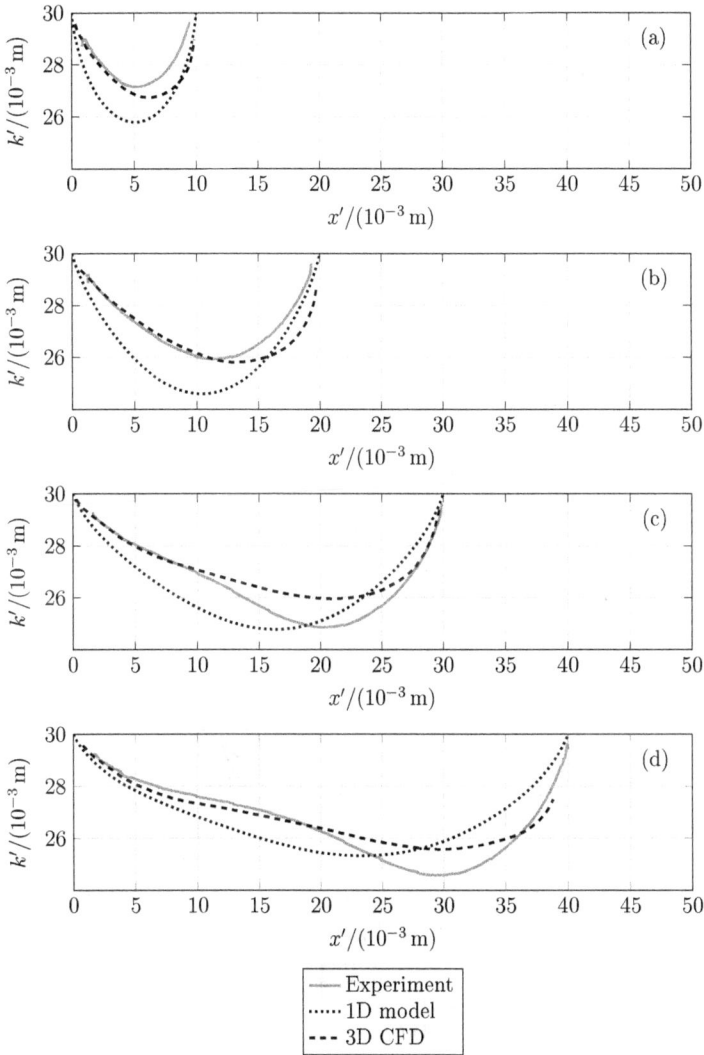

Figure 6.5: Comparison of free surface contours at respective experimental and numerical $Q'_{crit}$ for channel lengths in the wedge: (a) $l' = 10 \times 10^{-3}$ m, (b) $l' = 20 \times 10^{-3}$ m, (c) $l' = 30 \times 10^{-3}$ m, and (d) $l' = 40 \times 10^{-3}$ m.

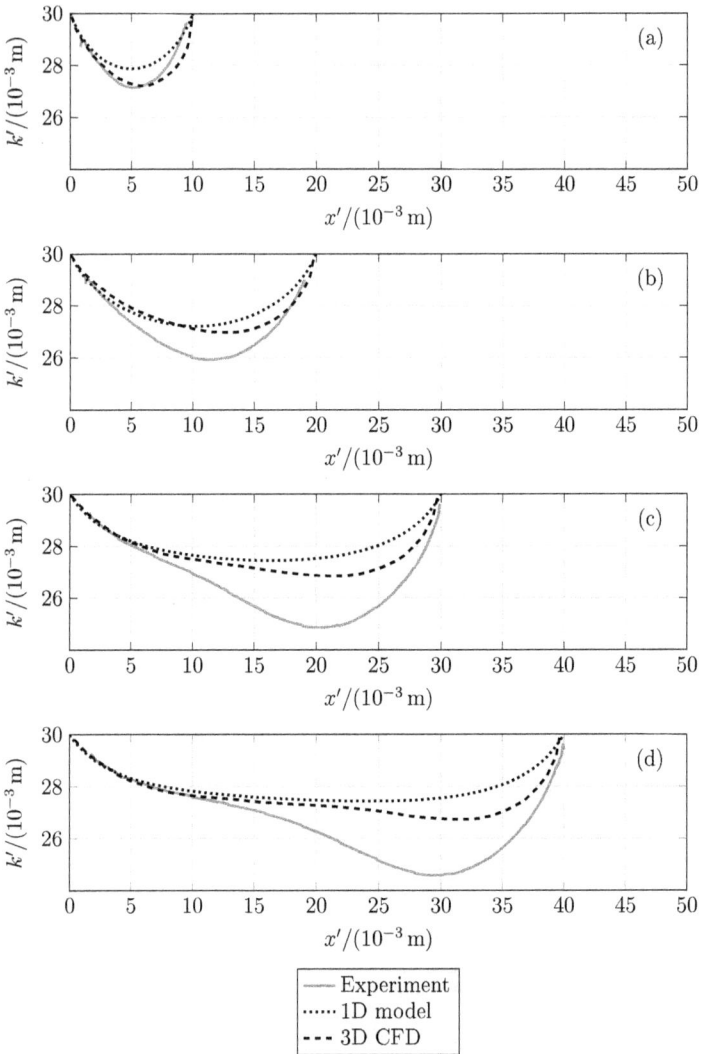

Figure 6.6: Comparison of free surface contours at $Q'_{crit,exp}$ for channel lengths in the wedge: (a) $l' = 10 \times 10^{-3}$ m, (b) $l' = 20 \times 10^{-3}$ m, (c) $l' = 30 \times 10^{-3}$ m, and (d) $l' = 40 \times 10^{-3}$ m.

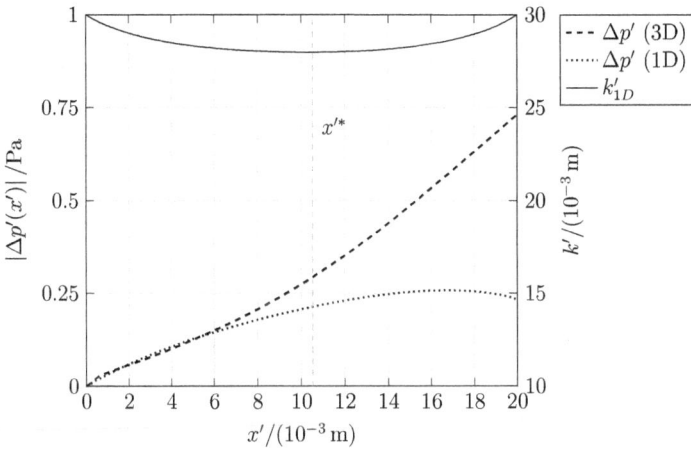

Figure 6.7: Pressure loss evolution within the test channel for $0.9\,Q'_{crit,exp}$ and $l' = 20 \times 10^{-3}\,\mathrm{m}$. The presented data is derived from numerical solutions with the 1D model and 3D CFD.

than $35 \times 10^{-3}\,\mathrm{m}$, the difference between all three results appears negligible. The relative error of the 1D simulations in this range is $\varepsilon_{1D} \approx 10\,\%$. For shorter channels with $l' < 35 \times 10^{-3}\,\mathrm{m}$, however, the 1D model deviates markedly from the experiment results and the relative error approaches $\varepsilon_{1D} \approx 30\,\%$. The difference between 1D and 3D computations in this regime indicates an additional pressure loss that is not included in the 1D model but is computed by the 3D CFD simulations. One possible explanation for the low pressure loss in the 1D simulations may lie in the fact that fully developed flow is assumed at the inlet of the test channel. As already discussed in section 4.3.3, the velocity profile is disturbed by the cannula of the BI and flow is actually not fully developed. Therefore, an additional pressure loss factor is present due to the development of the velocity profile according to entrance region theory.

Another likely candidate for neglected pressure loss in the 1D model is the fact that flow separation occurs at the free surface causing a recirculation zone downstream of $x'^*$. Such flow separation has been observed in previous capillary channel flow experiments (see section 1.2) and is clearly visible in the performed CFD simulations (see section 4.3). Figure 6.7 displays the pressure development along the channel for subcritical flow at $l' = 20 \times 10^{-3}$ m for both 1D and 3D computations. As discussed in section 2.2, the 1D model is based on streamtube theory and only viscous pressure loss at the walls is accounted for. Therefore, in the 1D model, convective pressure is regained at the outlet of the channel in accordance with streamtube theory. This basic assumption is contradicted by the 3D simulations. In the case displayed in figure 6.7, the pressure curves of the respective simulations begin to deviate around $l' = 6 \times 10^{-3}$ m and the convective pressure is not restored in the 3D computations. This supports the hypothesis that flow separation close to $x'^*$ leads to a significant pressure loss in the channel that is not accounted for in the 1D model because convective pressure is not regained when flow separation occurs. Therefore, the presented one-dimensional model most likely contains a model error that is most significant for short channels.

A comparison of the contour of the free surface at the respective critical flow rates is displayed in figure 6.5. How well the numerical results for the contour of the free surface coincides with the experiment result can be interpreted as a measure of the accuracy with which the pressure evolution in flow direction is computed. The pressure difference between the liquid and the surrounding gas along $x'$ is balanced by the mean curvature of the free surface at that location in accordance with the Young-Laplace equation (2.3) and the mean curvature is a function of $k'(x')$, the height of the contour as detailed in section 2.2. Therefore, an accurate simulation of $k'$ indicates that the pressure loss in the channel has been computed accurately. As can be seen in figure 6.5, the results of the experiments and the 3D simulations coincide quite well both at the inlet and along the length of the channel for the respective critical flow rates. On the other hand, the contour computed with the 1D model for $Q_{crit,1D}$, diverges significantly from the experiment results. For $l' <$

$30 \times 10^{-3}$ m this can be attributed to the fact that the critical flow rates found with the 1D model deviate significantly from the experiment results. The tendency of $Q'_{crit,1D} > Q'_{crit,exp}$, which is displayed in figure 6.4, is also visible in the pressure boundary condition at the inlet of the channel. According to equation (2.45), a higher flow rate results in a larger pressure difference $\sigma h'_0 = p'_a - p'_0$. The pressure boundary condition at $x'_0$ defines the angle at which the contour connects to the pinning edge according to the definition of $R'_2$ (compare equation (2.24); at $x'_0$, the first principal radius approaches infinity and therefore $1/R'_1 = 0$). The impact of the higher flow rate on the pressure boundary condition is significant in the comparison of the contours between 1D model and experiment in figure 6.5. In contrast, the critical flow rates determined in 3D simulations are quite close to those determined in the experiments and therefore the slope of the free surface in the vicinity of the inlet's pinning edge is quite accurate. It should be noted that the accuracy of the flow meter may also attribute to the divergence between 3D simulations and experiment because small changes in $Q'$ have larger effect on $k'(x')$ when $Q'$ is in the proximity of $Q'_{crit}$ as pointed out in chapter 5 and figures 5.4 and 5.5.

A further comparison of interface contours is displayed in figure 6.6, where numerical and experiment results are displayed for $Q'_{crit,exp}$ at $10 \times 10^{-3}$ m $\leq l' \leq 40 \times 10^{-3}$ m. Here the contours appear almost identical in the vicinity of the channel's inlet, which indicates that the pressure development within the entrance region is modeled very well in both CFD and 1D computations. Further along the channel, however, the contour of the experiment results deviates further and further from the numerical predictions and generally protrudes farther into the channel than the numerical counterparts.

Table 6.1: Dimensional and non-dimensional critical flow rates determined from experiments, one-dimensional model, and CFD simulations.

| $l'/\text{m}$ | $\tilde{l}$ | $Q'_{crit,exp}/(10^{-6}\,\text{m}^3\,\text{s}^{-1})$ | $Q_{crit,exp}$ | $Q'_{crit,1D}/(10^{-6}\,\text{m}^3\,\text{s}^{-1})$ | $Q_{crit,1D}$ | $Q'_{crit,3D}/(10^{-6}\,\text{m}^3\,\text{s}^{-1})$ | $Q_{crit,3D}$ |
|---|---|---|---|---|---|---|---|
| 0.0005 | 0.0000753 | 0.1551 | 2.5292 | 0.21034 | 3.430 | – | – |
| 0.0010 | 0.0001507 | 0.1250 | 2.0389 | 0.14428 | 2.353 | – | – |
| 0.0015 | 0.0002260 | 0.1084 | 1.7677 | 0.11592 | 1.890 | – | – |
| 0.0020 | 0.0003014 | 8.078 | 1.3173 | 9.9810 | 1.628 | – | – |
| 0.0025 | 0.0003767 | 6.899 | 1.1250 | 8.9470 | 1.459 | – | – |
| 0.0030 | 0.0004520 | 6.396 | 1.0429 | 8.2308 | 1.342 | – | – |
| 0.0035 | 0.0005274 | 5.802 | 0.9462 | 7.7046 | 1.256 | – | – |
| 0.0040 | 0.0006027 | 5.390 | 0.8789 | 7.2980 | 1.190 | – | – |
| 0.0050 | 0.0007534 | 4.955 | 0.8080 | 6.6975 | 1.092 | 5.200 | 0.8480 |
| 0.0060 | 0.0009041 | 4.663 | 0.7604 | 6.2510 | 1.019 | – | – |
| 0.0070 | 0.0010547 | 4.332 | 0.7064 | 5.8800 | 0.959 | – | – |
| 0.0080 | 0.0012054 | 4.162 | 0.6787 | 5.5627 | 0.907 | – | – |
| 0.0090 | 0.0013561 | 3.988 | 0.6503 | 5.2930 | 0.863 | – | – |
| 0.0100 | 0.0015068 | 3.851 | 0.6279 | 5.0611 | 0.825 | 4.100 | 0.6686 |
| 0.0110 | 0.0016575 | 3.734 | 0.6088 | 4.8610 | 0.793 | – | – |

Table 6.1: Dimensional and non-dimensional critical flow rates determined from experiments, one-dimensional model, and CFD simulations (continued).

| $l'/\mathrm{m}$ | $\tilde{l}$ | $Q'_{crit,exp}/(10^{-6}\,\mathrm{m^3\,s^{-1}})$ | $Q_{crit,exp}$ | $Q'_{crit,1D}/(10^{-6}\,\mathrm{m^3\,s^{-1}})$ | $Q_{crit,1D}$ | $Q'_{crit,OF}/(10^{-6}\,\mathrm{m^3\,s^{-1}})$ | $Q_{crit,3D}$ |
|---|---|---|---|---|---|---|---|
| 0.0120 | 0.0018081 | 3.617 | 0.5898 | 4.6860 | 0.764 | – | – |
| 0.0130 | 0.0019588 | 3.559 | 0.5804 | 4.5324 | 0.739 | – | – |
| 0.0140 | 0.0021095 | 3.465 | 0.5650 | 4.3961 | 0.717 | – | – |
| 0.0150 | 0.0022602 | 3.412 | 0.5564 | 4.2744 | 0.697 | 3.700 | 0.6034 |
| 0.0160 | 0.0024108 | 3.310 | 0.5397 | 4.1651 | 0.679 | – | – |
| 0.0170 | 0.0025615 | 3.310 | 0.5397 | 4.0660 | 0.663 | – | – |
| 0.0180 | 0.0027122 | 3.210 | 0.5235 | 3.9770 | 0.649 | – | – |
| 0.0190 | 0.0028629 | 3.177 | 0.5181 | 3.8953 | 0.635 | – | – |
| 0.0200 | 0.0030136 | 3.109 | 0.5069 | 3.8210 | 0.623 | 3.400 | 0.5544 |
| 0.0210 | 0.0031642 | 3.110 | 0.5072 | 3.7530 | 0.612 | – | – |
| 0.0220 | 0.0033149 | 3.077 | 0.5018 | 3.6900 | 0.602 | – | – |
| 0.0230 | 0.0034656 | 3.011 | 0.4910 | 3.6330 | 0.592 | – | – |
| 0.0240 | 0.0036163 | 3.011 | 0.4910 | 3.5800 | 0.584 | – | – |
| 0.0250 | 0.0037670 | 2.957 | 0.4822 | 3.5340 | 0.576 | 3.200 | 0.5218 |

Table 6.1: Dimensional and non-dimensional critical flow rates determined from experiments, one-dimensional model, and CFD simulations (continued).

| $V'/\mathrm{m}$ | $\tilde{I}$ | $Q'_{crit,exp}/(10^{-6}\,\mathrm{m^3\,s^{-1}})$ | $Q_{crit,exp}$ | $Q'_{crit,1D}/(10^{-6}\,\mathrm{m^3\,s^{-1}})$ | $Q_{crit,1D}$ | $Q'_{crit,OF}/(10^{-6}\,\mathrm{m^3\,s^{-1}})$ | $Q_{crit,3D}$ |
|---|---|---|---|---|---|---|---|
| 0.0260 | 0.0039176 | 2.977 | 0.4855 | 3.4860 | 0.568 | – | – |
| 0.0270 | 0.0040683 | 2.936 | 0.4788 | 3.4450 | 0.562 | – | – |
| 0.0280 | 0.0042190 | 2.911 | 0.4747 | 3.4060 | 0.555 | – | – |
| 0.0290 | 0.0043697 | 2.911 | 0.4747 | 3.3700 | 0.550 | – | – |
| 0.0300 | 0.0045203 | 2.905 | 0.4737 | 3.3370 | 0.544 | 3.100 | 0.5055 |
| 0.0310 | 0.0046710 | 2.911 | 0.4747 | 3.3059 | 0.539 | – | – |
| 0.0320 | 0.0048217 | 2.878 | 0.4693 | 3.2770 | 0.534 | – | – |
| 0.0330 | 0.0049724 | 2.836 | 0.4625 | 3.2500 | 0.530 | – | – |
| 0.0340 | 0.0051231 | 2.811 | 0.4584 | 3.2250 | 0.526 | – | – |
| 0.0350 | 0.0052737 | 2.848 | 0.4644 | 3.2010 | 0.522 | 3.100 | 0.5055 |
| 0.0360 | 0.0054244 | 2.811 | 0.4584 | 3.1800 | 0.519 | – | – |
| 0.0370 | 0.0055751 | 2.811 | 0.4584 | 3.1580 | 0.515 | – | – |
| 0.0380 | 0.0057258 | 2.778 | 0.4530 | 3.1380 | 0.512 | – | – |
| 0.0390 | 0.0058764 | 2.737 | 0.4462 | 3.1198 | 0.509 | – | – |

Table 6.1: Dimensional and non-dimensional critical flow rates determined from experiments, one-dimensional model, and CFD simulations (continued).

| $l'$/m | $\tilde{l}$ | $Q'_{crit,exp}/(10^{-6}\,\mathrm{m}^3\,\mathrm{s}^{-1})$ | $Q_{crit,exp}$ | $Q'_{crit,1D}/(10^{-6}\,\mathrm{m}^3\,\mathrm{s}^{-1})$ | $Q_{crit,1D}$ | $Q'_{crit,OF}/(10^{-6}\,\mathrm{m}^3\,\mathrm{s}^{-1})$ | $Q_{crit,3D}$ |
|---|---|---|---|---|---|---|---|
| 0.0400 | 0.0060271 | 2.754 | 0.4491 | 3.1020 | 0.506 | 3.000 | 0.4892 |
| 0.0410 | 0.0061778 | 2.712 | 0.4422 | 3.0855 | 0.503 | – | – |
| 0.0420 | 0.0063285 | 2.712 | 0.4422 | 3.0690 | 0.500 | – | – |
| 0.0430 | 0.0064792 | 2.712 | 0.4422 | 3.0400 | 0.498 | – | – |
| 0.0440 | 0.0066298 | 2.712 | 0.4422 | 3.0260 | 0.496 | – | – |
| 0.0450 | 0.0067805 | 2.705 | 0.4411 | 3.0120 | 0.493 | 3.000 | 0.4892 |
| 0.0460 | 0.0069312 | 2.712 | 0.4422 | 2.9990 | 0.491 | – | – |
| 0.0470 | 0.0070819 | 2.712 | 0.4422 | 2.9990 | 0.489 | – | – |
| 0.0480 | 0.0072325 | 2.675 | 0.4363 | 2.9860 | 0.487 | 2.900 | 0.4729 |

## 6.3    Subcritical Single-Phase Flow

As stated above, the contour of the interface is very sensitive in the immediate vicinity of $Q'_{crit}$ and small variations in $Q'$ may have large effects on the equilibrium shape of the interface. Therefore an additional comparison of interface contours is provided in figures 6.8 and 6.9 for subcritical flow rates $0.9\,Q'_{crit,exp}$ and $0.8\,Q_{crit,exp}$ respectively. The channel lengths chosen to be depicted are in the range of $10 \times 10^{-3}\,\text{m} \leq l' \leq 40 \times 10^{-3}\,\text{m}$. In contrast to the results at $Q'_{crit}$, the computed results of both CFD and one-dimensional model are in excellent agreement with the experiment results and the contour lines of the numerical results coincide well with the results form the experiments on the ISS.

## 6.4    Two-Phase Critical Flow

Critical flow rates for two-phase flow were determined experimentally for the channel lengths $l' = \{10, 15, 20, 25, 30\} \times 10^{-3}\,\text{m}$ and with averaged gas flow rates $Q'_G = \{0.1, 0.2, 0.3, 0.4\} \times 10^{-6}\,\text{m}^3\,\text{s}^{-1}$. Frequency and duty cycle of valve C2 were constant for all experiments with $f'_{C2} = 2\,\text{s}^{-1}$ and $t'_{C2} = 0.15\,\text{s}$ respectively. The average gas flow rate injected into the channel was adjusted via the pressure difference between the gas reservoir K3 and $p'_0$ in accordance with Weislogel et al. [64]. In some cases images were recorded using the HSHRC. Figure 6.10 displays a representative snapshot of the video data acquired during an experiment run. The figure displays flow from left to right at subcritical flow with the stable interface visible in the upper portion of the image. The location of the vertex (VE) is indicated in the lower portion of the image. A mono-disperse bubble stream is visible in the centre of the channel. In the lower left hand corner the location of the BI is indicated by the gaseous jet caused during a bubble injection cycle. The bubbles were generally found to be of constant volume and spherical shape once they had risen to an appropriate height in the channel to accommodate their diameter. It should be noted that bubble rise seen in this figure is not due to buoyancy but rather capillary effects [64].

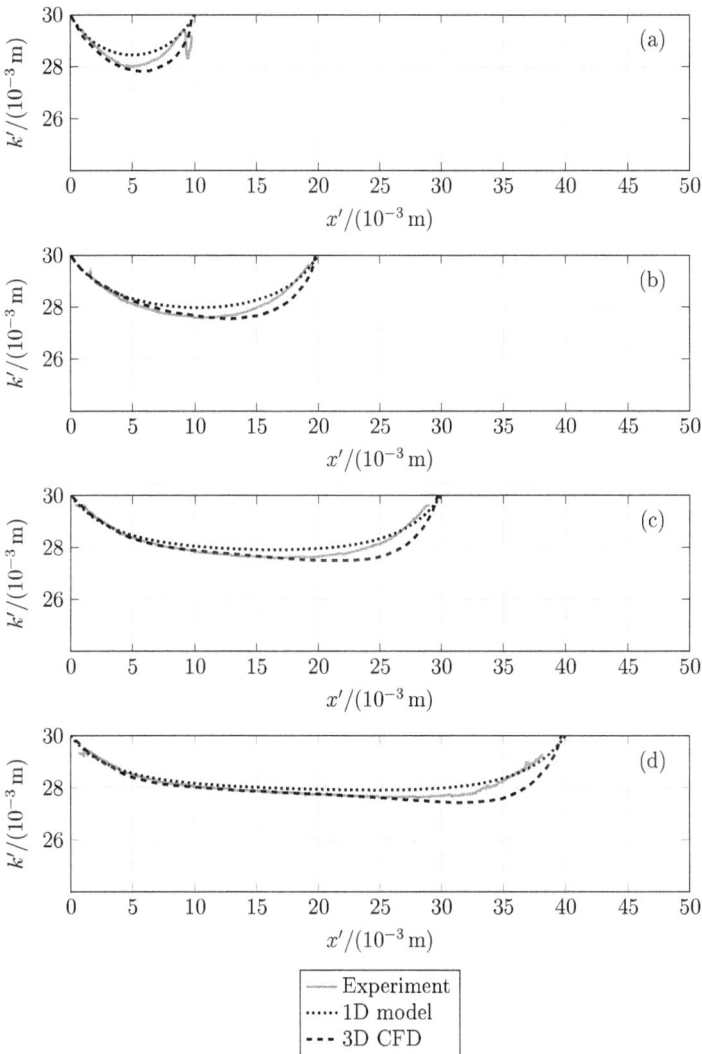

Figure 6.8: Comparison of free surface contours at $0.9\,Q'_{crit,exp}$ for channel lengths: (a) $l' = 10 \times 10^{-3}\,\mathrm{m}$, (b) $l' = 20 \times 10^{-3}\,\mathrm{m}$, (c) $l' = 30 \times 10^{-3}\,\mathrm{m}$, and (d) $l' = 40 \times 10^{-3}\,\mathrm{m}$.

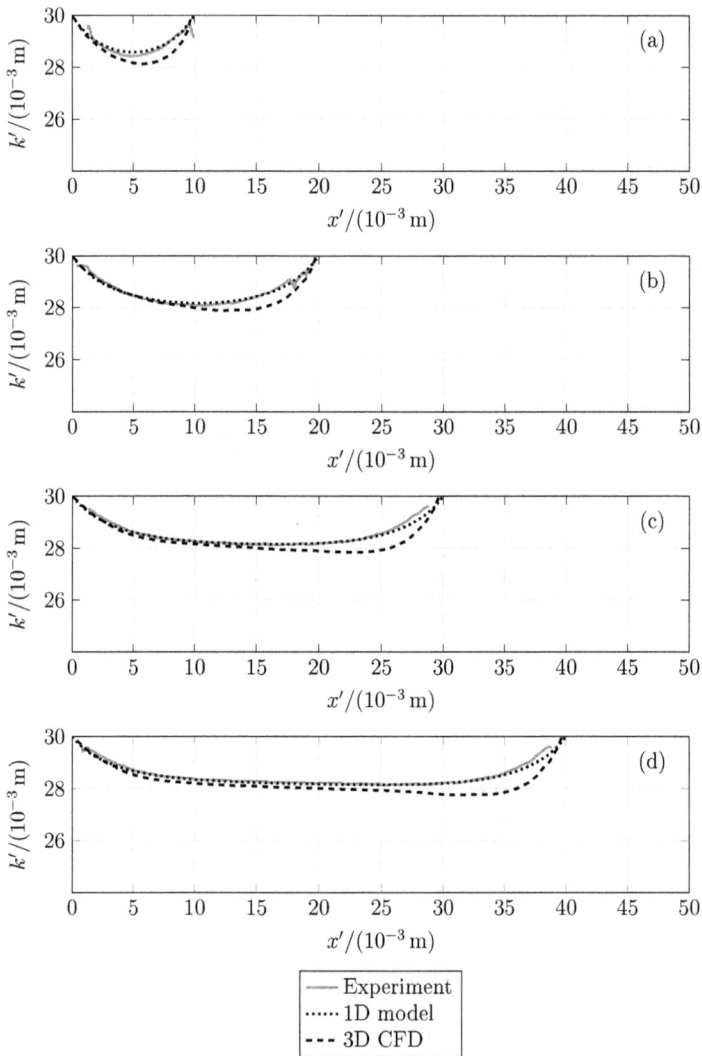

Figure 6.9: Comparison of free surface contours at $0.8\,Q'_{crit,exp}$ for channel lengths: (a) $l' = 10 \times 10^{-3}\,\mathrm{m}$, (b) $l' = 20 \times 10^{-3}\,\mathrm{m}$, (c) $l' = 30 \times 10^{-3}\,\mathrm{m}$, and (d) $l' = 40 \times 10^{-3}\,\mathrm{m}$.

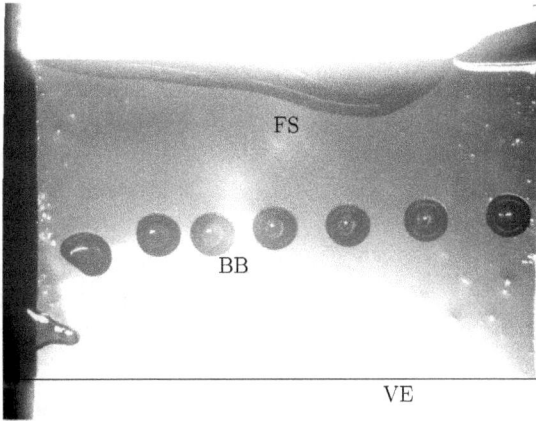

Figure 6.10: Representative snapshot of two-phase flow experiment with bubble injection. Flow is from left to right. A mono-disperse stream of bubbles (BB) is introduced into the TC just above the vertex (VE). Bubbles are considered to be approximately spherical and of equal size.

Each bubble injection caused a disturbance of the slope of the interface at the channel's inlet which led to waves on the interface. The waves may be caused by a periodic alteration of the pressure boundary condition at the inlet of the test channel as discussed in section 2.3. Periodic stepwise variation of the pressure boundary condition would cause the slope to decrease and increase with bubble injection thus generating waves on the interface. For flow rates in the vicinity of $Q'_{crit,2P}$, the interface displayed a periodic semi-transient behaviour that may be described as supercritical choking but sudden restabilization caused by bubble injection before a bubble is choked into the channel. In this way the interface would swing back and forth, periodically bending into the channel and then retracting back to a stable position but never actually collapsing and ingesting gas.

In the experiments, gas injection was found to have a stabilizing effect on the free surface and enhanced the critical flow rate. A further noteworthy observation during the experiments was that the interface collapsed when $f'_{C2} < f'_{ing}$ for a given

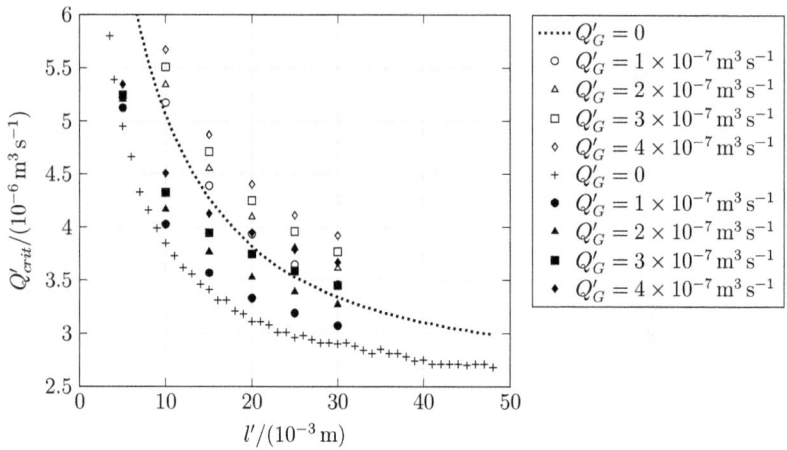

Figure 6.11: Comparison between numerically and experimentally determined critical flow rates in two-phase flow for various gas injection flow rates. Open marks indicate model results, filled marks indicate experiment results.

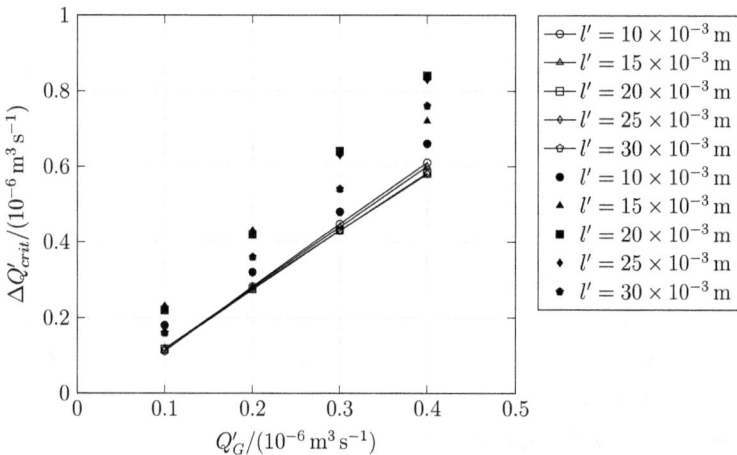

Figure 6.12: Comparison of numerical and experimental results for $\Delta Q'_{crit,2P}$. Filled markers denote experiment results, empty markers with lines are numerical results.

channel length and flow rate $Q'$. The enhancement of $Q'_{crit}$ exceeds the volumetric
gas flow rate of bubble injection. A direct comparison between the experimentally
and numerically determined absolute values for critical two-phase flow are displayed
in figure 6.11. Due to the difference between the single-phase critical flow rates of
both studies, the absolute values cannot be expected to coincide. Consequently,
the enhancement of the critical flow rate, $\Delta Q'_{crit}$ is plotted in figure 6.12 for both
cases for a more appropriate comparison. This means that the absolute values of
the critical flow rates are not necessarily of interest for this comparison, but rather
the effect the gas injection has on the critical flow rate enhancement. A comparison
of the results shows that the model under-predicts the enhancement of the critical
flow rate. Also, the numerical results obtained with the modified 1D model display
hardly any dependency on $l'$ whereas the experiment results diverge but not with
any particularly apparent trend.

Errors between experiment and model may arise from the difficulty of main-
taining constant bubble sizes during the experiments. In addition, the accuracy of
the experiment setup may have an impact on the observed results considering the
orders of magnitude that is used to display experiment data ($\Delta Q'_{crit}$ is of similar
magnitude as the flow meter accuracy discussed in chapter 3). Nonetheless, considering the simplicity of the model used and the number of assumptions it is based
on, the results of the modified 1D model display the same trend as the experiment
results and may be used for approximate predictions of the two-phase critical flow
rate in the parameter range investigated. Interestingly, the viscous pressure loss
in the channel appears to be well-modelled using the single-phase friction factor as
discussed in section 2.3. As discussed in chapter 2, an increased viscous pressure
loss might have been expected. But an increased friction factor would further reduce $\Delta Q'_{crit}$ making the results coincide less with the experiment results. The good
agreement of the chosen viscous model may be attributed to the fact that $\beta \leq 0.12$
for the investigated cases. Also, the injected bubbles are mono-disperse and remain
relatively small in comparison to $A'_0$ ($0.164 \times 10^{-4}\,\mathrm{m}^2 \leq A'_{BB} \leq 0.435 \times 10^{-4}\,\mathrm{m}^2$;
$0.13 \leq A'_{BB}/A'_0 \leq 0.33$) and might therefore be assumed to have little impact on

the flow field of the surrounding liquid. Instead, the liquid transports the bubbles according to the velocity field within the test channel. This coincides with the work of Weislogel et al. [64], who in the same experiment setup observed that the injected bubbles approach velocities expected of the liquid velocity profile at their respective centroids.

## 6.5   Supercritical Bubble Ingestion

Choking and bubble ingestion occurs in steady flow when $Q'_{crit}$ is exceeded. Experiments were performed at supercritical flow rates $\Delta Q'_{sc}$ without the use of the bubble injector to determine whether the gas ingestion behaviour changes when the flow rate is increased further. The supercritical flow rate is defined for $Q' > Q'_{crit}$ as follows:

$$Q' = Q'_{crit} + \Delta Q'_{sc} \quad . \tag{6.2}$$

Bubble ingestion frequencies $f'_{ing}$ and bubble volumes $V'_b$ were determined for selected channel lengths and only in the groove and wedge channels. Multiple ingestion events were recorded for each data point and the presented results of $f'_{ing}$ and $V'_b$ are averaged values. An average gas ingestion flow rate during choking is defined simply as $Q'_{ch} = V'_b f'_{ing}$.

The mean diameter $\bar{D}'_b$ of all ingested bubbles for each $V'_b$ was measured in the $x'z'$-plane (compare figures 3.3 and 3.4). Non-circularity of bubbles in the $x'z'$-plane is accounted for by determining a mean diameter for each individual bubble by measuring $D'_b$ at multiple angles $\psi$. All ingested bubble diameters were found to be larger than the channel's width, which means that they cannot be spherical but are confined between the plates and have a shape that resembles a cheese wheel. As the test liquid is perfectly wetting, the outer rim of the bubbles is curved and tangential to the test channel's walls. As the recorded images only show the $x'z'$-plane, the three-dimensional shape of the bubbles can only be assumed. The shape of the bubble that is used to calculate $V'_b$ is shown on the left hand side of figure 6.13. The

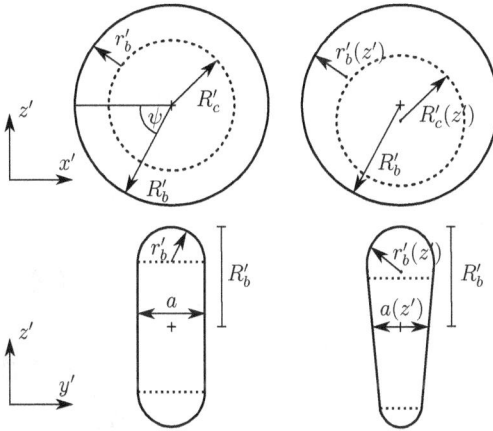

Figure 6.13: Assumed shape of the ingested bubbles in the $x'z'$-plane (top) and the $y'z'$-plane (bottom). In EU1 the channel's walls are parallel and thus the bubble has parallel interfaces (left). The tilt of the walls of the wedge in EU2 is taken into consideration for the assumed shape of the ingested bubbles (right).

volume of the bubble in EU1 can then be calculated as the sum of the outer half of a ring torus and an inner circular cylinder as indicated in figure 6.13 with the total volume $V_b'$ given as

$$V_b' = \pi^2 r_b'^3 \left( \frac{R_b'}{r_b'} + \frac{4}{3\pi} - 1 \right) + 2\pi r_b' \left( R_b' - r_b' \right)^2 \quad , \tag{6.3}$$

where $R_b' = 0.5D_b'$ is the measured radius of the bubble and $r_b'$ is assumed to be equal to $0.5a$.

In EU2 the ingested bubbles are deformed according to the tilt in the channel's walls. The assumed shape is shown on the right hand side of figure 6.13. The position of the bubble's centre is determined in addition to its diameter because the width of the channel is a function of the height $z'$ within the channel (compare figure 3.3). An estimate of the bubble's volume is given by using equation (6.3) with the channel's width $a(z')$ at the height of the bubble's centre.

The results are summarized in figures 6.14, 6.15, and 6.16 for both the groove and the wedge channels. Both channels display similar behaviour when the flow rate

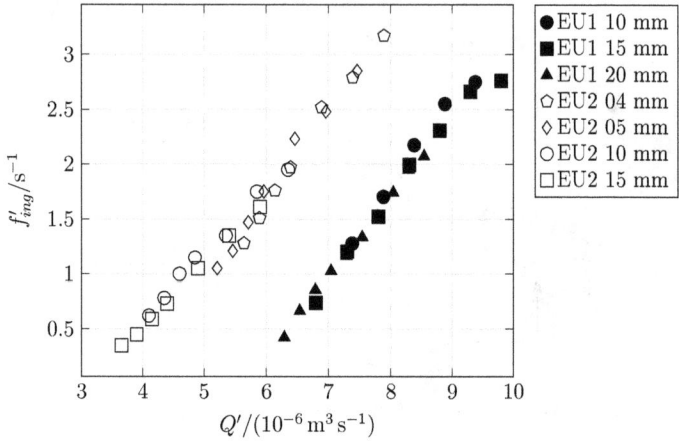

Figure 6.14: Bubble ingestion frequencies for various channel lengths observed in experiments at supercritical flow rates in EU1 and EU2.

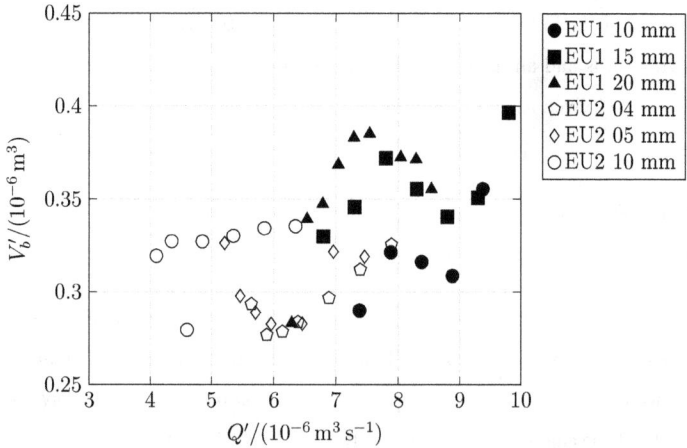

Figure 6.15: Averaged volumes of ingested bubbles for various channel lengths observed in experiments at supercritical flow rates in EU1 and EU2.

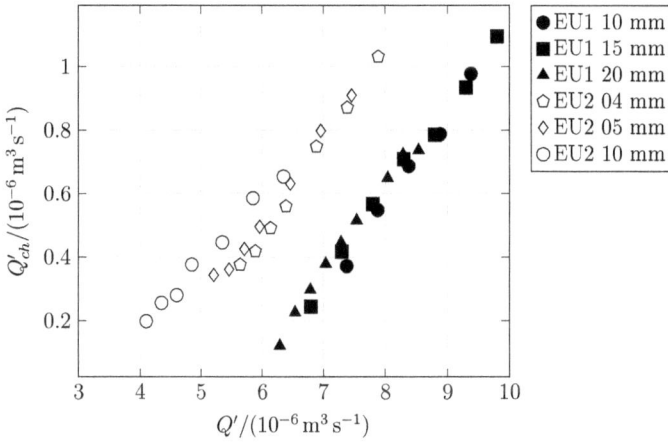

Figure 6.16: Time-averaged gas flow rate via ingestion during choking at various supercritical flow rates and for EU1 and EU2.

is increased further than $Q'_{crit}$. The ingestion frequency rises when the flow rate is increased. This dependency appears to be linear. Furthermore, it is observed that shorter channels produce higher frequencies than longer channels at identical $\Delta Q'_{sc}$. The volume of the ingested bubbles appears to remain within a small range (maximum variation of $V'_b$ is less than $0.12 \times 10^{-6}\,\mathrm{m}^3$ for EU2 and even lower for EU1) and displays no immediate dependency on the flow rate. Further work is required to fully understand the behaviour of the free surface and the gas ingestion process in supercritical flow and the trends that are shown here suggest that modelled predictions of $V'_b$ and $f'_{ing}$ may be possible for other channel lengths and geometries. An attempt to predict the behaviour of the free surface at supercritical flow rates in EU1 is undertaken in chapter 7.

# Chapter 7

# Bubble Ingestion Model

In the supercritical flow regime, the ingested bubbles display highly repeatable values for bubble volume and ingestion frequency (compare section 6.5). The ingested bubble volume appears to vary negligibly between the modified parameters and is therefore considered to be independent of the parameters $l'$ and supercritical flow rate $Q' = Q'_{crit} + \Delta Q'_{sc}$. The ingestion frequency, $f'_{ing}$, displays an evident dependency both on the channel's length and on $Q'$. In fact, for constant $l'$, the ingestion frequency appears to increase linearly with $\Delta Q'_{sc}$. Based on observations made throughout performing and evaluating the experiments at $Q' > Q'_{crit}$, a simple time-dependent model is defined to predict bubble growth. The presented model is defined for and applies only to the groove shaped channel in EU1 (compare chapter 3).

The basic principle of the proposed model is best summarized using snapshots from the unsteady behaviour of the free surface during supercritical flow. The regions of interest of three representative frames recorded during a supercritical flow experiment are displayed in figure 7.1. Each subfigure displays both the observed contour of the free surface and the model's predicted bubble size and location of $x'^*$ and $k'^*$ for the individual time step. In subfigure (a), a bubble has just been ingested into the flow and the interface has retracted. This is the starting point for the model. Subfigure (b) presents a typical snapshot of the interface about halfway

through the bubble growth phase. Subfigure (c) displays the contour of the free
surface immediately before the growing bubble detaches and is ingested into the flow.
This marks the end of a single bubble ingestion phase and the process repeats itself
until the interface is restabilized by decreasing $Q'$ sufficiently. The presented model
is used to predict $x'^*$ and $k'^*$ as functions of time (see figure 7.1). Bubble growth is
modelled as a growing cylindrical bubble with rounded edges that is transported in
flow direction throughout the growth phase (compare dashed circles in figure 7.1).
In section 7.2, predicted $x'^*(t)$ and $k'^*(t)$ are compared with experiment results for
the case displayed in figure 7.1. While the model is used to predict the bubble
growth development over time *a priori*, empirical data on average bubble volume is
required to determine the ingestion frequency.

# 7.1   Model Assumptions

In the following, a number of assumptions will be declared upon which the presented
model is based. These assumptions are based on observations during the experiments
and from studying numerous videos and still frames of the bubble growth process
such as those displayed[1] in figure 7.1.

## 7.1.1   Velocity limit drives bubble growth

The model for bubble growth in supercritical flow is based primarily on the assumption that flow velocity within the open capillary channel cannot exceed the capillary
wave speed $v'_{ca}$. According to Rosendahl et al. [46], increasing the pump speed (i.e.
increasing the forced flow rate at the channel's outlet) generates a pressure disturbance that propagates upstream along the flow path. Within a closed duct with solid
walls, such a pressure disturbance travels at the speed of sound within the fluid. At
low Bond numbers and when the solid wall is removed leaving the flow bounded by

---

[1]The displayed experiment results are taken from frames 414 ($t' = 0$ s), 440 ($t' = 0.43$ s), and
464 ($t' = 0.82$ s) of GR01032.

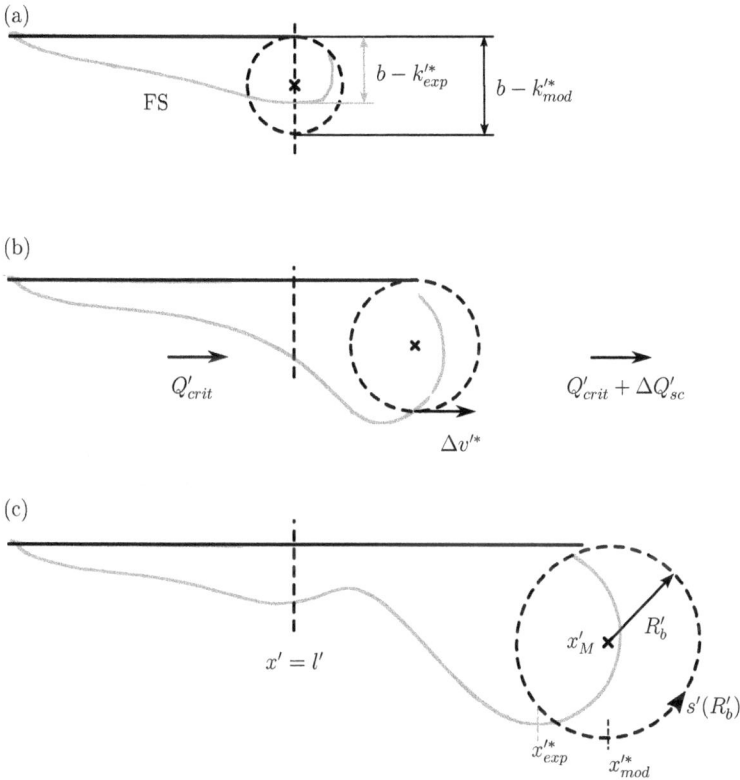

Figure 7.1: Observations of supercritical flow experiments (grey outline) upon which assumptions for the bubble ingestion model (dashed circle) are based. From an initial position **(a)** immediately after a prior pinch-off event, bubble growth **(b)** is driven by the relative velocity $\Delta v'^*$ at the smallest cross-section $x'^*$, where $k' = k'^*$. As the bubble grows, the curvature of the interface downstream of $x'^*$ is approximately constant. When the bubble has grown large enough and traveled far enough, a shrinking neck is formed **(c)** and ultimately the bubble pinches off into the surrounding flow. The overlaid circles represent the model's prediction at the respective experiment time steps.

a free surface, as is the case in open capillary channel flow, the pressure disturbance travels at the capillary wave speed $v'_{ca} = (-A' \mathrm{d}h'/\mathrm{d}A')^{1/2}$. As the flow rate is increased further, the local average velocity $v'^*$ in the channel's smallest cross-section $A'^*$ at $x' = x'^*$ also increases and reaches an upper limit at $v'^* = v'_{ca}$ analogously to the velocity limit in compressible flow through a converging-diverging nozzle, which is limited by the speed of sound (density wave speed). This limit is also described in previous chapters as the critical flow rate $Q'_{crit}$, which can be defined as

$$Q'_{crit} = v'_{ca} A'^*_{crit} \quad , \tag{7.1}$$

where $A'^*_{crit}$ is the smallest cross-section in the channel which is located at $x'^*$ at $Q' = Q'_{crit}$.

For steady flow at $Q' \leq Q'_{crit}$ the interface adopts a stable concave shape bounded between $x' = 0\,\mathrm{m}$ and $x' = l$ and $x'^*$ is constant. When the flow rate is increased to exceed $Q'_{crit}$ the interface is forced to move in flow direction because the local average velocity $v'^*$ has already reached its limit $v'_{ca}$. But, because the free surface is a flexible boundary, $v'^*$ can increase further if $x'^*$ travels downstream at the boundary velocity $\Delta v'^*$ with

$$\Delta v'^* = \frac{x'^*(t' + \Delta t') - x'^*(t')}{\Delta t'} = v'^* - v'_{ca} \quad . \tag{7.2}$$

Thus, the velocity of $x'^*$ can be determined using equations (7.1) and (7.2) and the values for $v'_{ca}$ and $A'^*$ that are determined using the 1D model for steady flow in a rectangular groove (compare Bronowicki et al. [9]).

### 7.1.2   Constant Curvature

In accordance with the continuity equation, the highest velocity in the test channel is expected to be located in the cross-section with the lowest area. This location is defined as $x'^*$. The curvature of the free surface is assumed constant for $x' > x'^*$ during bubble ingestion and the width of the ingested bubble is approximately equal to the channel width. Observations of the ingestion process suggest that the local curvature of the interface assumes a time-dependent constant value downstream of the smallest cross-sectional area $A'^*$. Additionally, the width of the bubble is constant throughout the ingestion process and is constrained by the channel width $a$. Based on these observations the ingestion process is modelled as the growth of a circular cylindrical bubble with a time dependent outer radius $R'_b$ and with rounded edges with a constant inner radius of $r'_b = a/2$ as displayed in figure 6.13. The arc length of the outer perimeter of the modelled bubble is defined as $s' = 2\pi R'_b$. As the bubble grows and moves into the channel, a neck is formed whose circumference shrinks as the bubble progresses farther into the channel and ultimately detaches. The model neglects the growth of the neck zone and is therefore likely to over predict the radius of the growing bubble.

### 7.1.3 Arc Length Growth Rate

Combining the assumptions stated in sections 7.1.1 and 7.1.2 yields the growth rate of the ingested bubble. The translation of the interface leads to an increased arc length of the semi-circular interface downstream of $x'^*$. The additional arc length is defined as follows:

$$\Delta s' = \Delta v'^* \Delta t' \quad , \tag{7.3}$$

where $\Delta t'$ is a discrete time step that is equal to the time interval between frames recorded by the high speed camera in the experiments. Arc length is increased both at the top and the bottom of the circular interface at $x' = x'^*$ (compare figure 7.1). The increased arc length yields a larger radius for the ingested bubble where $R'_b(t' + \Delta t') = R'_b(t') + \Delta R'_b$ and the difference between radii of two time steps is defined as:

$$\Delta R'_b = \frac{\Delta s'}{2\pi} \quad . \tag{7.4}$$

Furthermore, it is assumed that the top edge of the bubble is located at $z' = b/2$ throughout the ingestion phase and the lowest edge of the bubble is then located at $k'^* = b/2 - 2R'_b$.

Table 7.1: Overview of supercritical experiments in rectangular groove channel that are compared to the presented supercritical model.

| EXP No. | $l'/(10^{-3}\,\mathrm{m})$ | $\Delta Q'_{sc}/(10^{-6}\,\mathrm{m}^3\,\mathrm{s}^{-1})$ |
|---------|------|-----|
| GR1022  | 10   | 0.5 |
| GR1026  | 10   | 1.0 |
| GR1027  | 10   | 1.5 |
| GR1029  | 10   | 2.0 |
| GR1031  | 15   | 0.5 |
| GR1032  | 15   | 1.0 |
| GR1033  | 15   | 1.5 |
| GR1034  | 15   | 2.0 |
| GR1051  | 20   | 0.5 |
| GR1053  | 20   | 1.0 |
| GR1055  | 20   | 1.5 |
| GR1059  | 20   | 2.0 |

## 7.1.4   Initial Conditions

In the model, the initial bubble radius is defined as the largest possible radius for spherical bubble constrained within the channel geometry and connected to the downstream pinning edge. A visual representation of this initial condition is sketched in figure 7.1 (a). For the examined channel geometry (groove), the initial radius is then defined as $R'_b(t' = 0) = a/2$. As displayed in figure 7.1, the initial position of the modelled bubble is such that the bubble's centre is located at $x' = l'$.

# 7.2   Model Results

The results of the modelled supercritical bubble ingestion are the evolution of $x'^*$ and $k'^*$ over time which are displayed in figures 7.2 to 7.13 where they are compared with the evaluated contour evolution in experiments with supercritical flow rates. Both channel length and supercritical flow rate were varied in experiments and model. The complete data set that was used to compare the model is displayed in table 7.1. The plots show $k'^*$ and $x'^*$ as a function of time where $\Delta x'$ is the distance that $x'^*$ travels throughout the generation of a single bubble. The general trend of the displayed time dependent variables coincides very well with the data evaluated from the experiment. It should be noted that only individual ingestion events are compared and no experimental data has been averaged. As shown in section 6.5, frequency and volume of bubble ingestion are quite regular, but even small differences become apparent when comparing $x'^*(t')$ and $k'^*(t')$ between two ingestion events.

Unfortunately, while the model appears to describe the choking process quite well, as yet it has no end condition to determine when a bubble pinches off and the ingestion process is repeated. As can be seen in some of the plotted experiment results, the end of the bubble growth is clearly visible as a distinct divergence from the general trend of $x'^*(t')$ and $k'^*(t')$. This divergence is particulary visible in figures 7.8 and 7.13, where the divergence occurs at around $t' = 0.65\,\mathrm{s}$ and $t' = 0.6\,\mathrm{s}$ respectively. An empirical method would be possible by using the average bubble volume evaluated from supercritical flow experiments as a limiting factor in the model. Implementation of the ingestion model to the wedge-shaped channel would be possible by adapting the input parameters $A'^*$ and $v'_{ca}$ to the according values from the steady flow model presented in chapter 2.2. In addition, the confinement of the growing ingested bubble would have to be accounted for as it grows larger and travels deeper into the triangular cross-section.

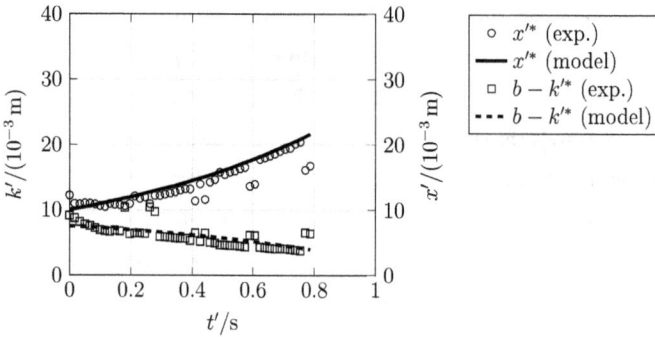

Figure 7.2: Time evolution of $x'^*$ and $k'^*$ during choking for experiment GR01022 with $l' = 10 \times 10^{-3}$ m and $\Delta Q'_{sc} = 0.5 \times 10^{-6}\,\mathrm{m^3\,s^{-1}}$. Comparison of supercritical model and experiment results. Outliers may occur due to difficulty of postprocessing the moving interface.

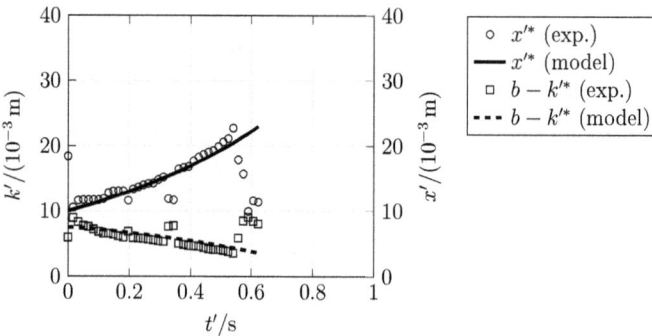

Figure 7.3: Time evolution of $x'^*$ and $k'^*$ during choking for experiment GR01026 with $l' = 10 \times 10^{-3}$ m and $\Delta Q'_{sc} = 1 \times 10^{-6}\,\mathrm{m^3\,s^{-1}}$. Comparison of supercritical model and experiment results. Outliers may occur due to difficulty of postprocessing the moving interface.

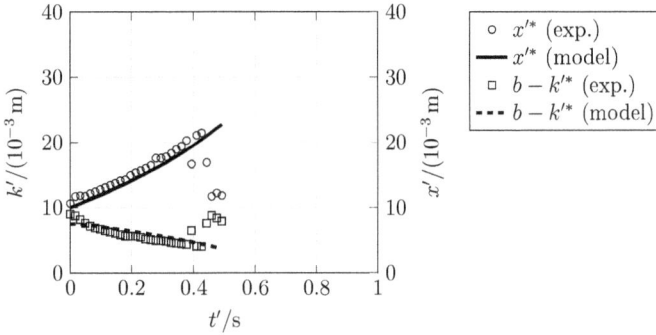

Figure 7.4: Time evolution of $x'^*$ and $k'^*$ during choking for experiment GR01027 with $l' = 10 \times 10^{-3}$ m and $\Delta Q'_{sc} = 1.5 \times 10^{-6}$ m$^3$ s$^{-1}$. Comparison of supercritical model and experiment results. Outliers may occur due to difficulty of postprocessing the moving interface.

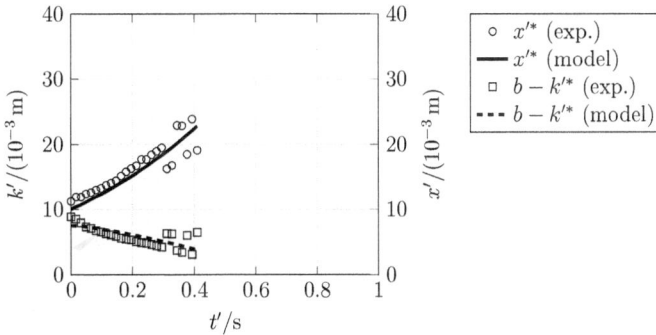

Figure 7.5: Time evolution of $x'^*$ and $k'^*$ during choking for experiment GR01029 with $l' = 10 \times 10^{-3}$ m and $\Delta Q'_{sc} = 2 \times 10^{-6}$ m$^3$ s$^{-1}$. Comparison of supercritical model and experiment results. Outliers may occur due to difficulty of postprocessing the moving interface.

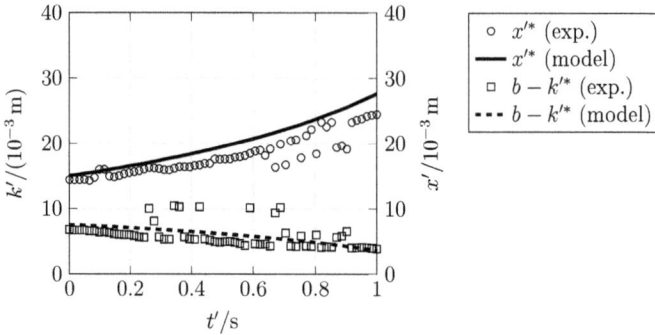

Figure 7.6: Time evolution of $x'^*$ and $k'^*$ during choking for experiment GR01031 with $l' = 15 \times 10^{-3}\,\text{m}$ and $\Delta Q'_{sc} = 0.5 \times 10^{-6}\,\text{m}^3\,\text{s}^{-1}$. Comparison of supercritical model and experiment results. Outliers may occur due to difficulty of postprocessing the moving interface.

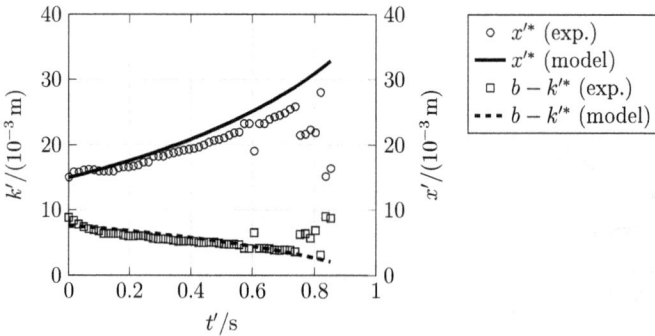

Figure 7.7: Time evolution of $x'^*$ and $k'^*$ during choking for experiment GR01032 with $l' = 15 \times 10^{-3}\,\text{m}$ and $\Delta Q'_{sc} = 1 \times 10^{-6}\,\text{m}^3\,\text{s}^{-1}$. Comparison of supercritical model and experiment results. Outliers may occur due to difficulty of postprocessing the moving interface.

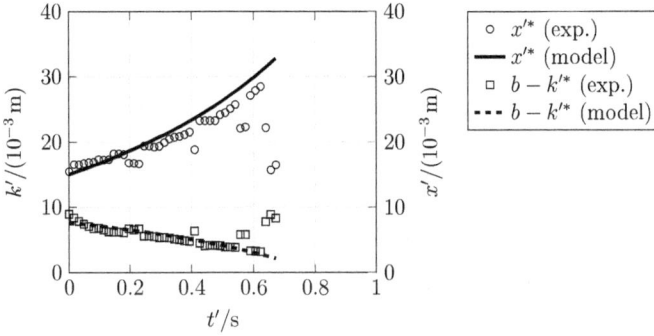

Figure 7.8: Time evolution of $x'^*$ and $k'^*$ during choking for experiment GR01033 with $l' = 15 \times 10^{-3}$ m and $\Delta Q'_{sc} = 1.5 \times 10^{-6}$ m$^3$ s$^{-1}$. Comparison of supercritical model and experiment results. Outliers may occur due to difficulty of postprocessing the moving interface.

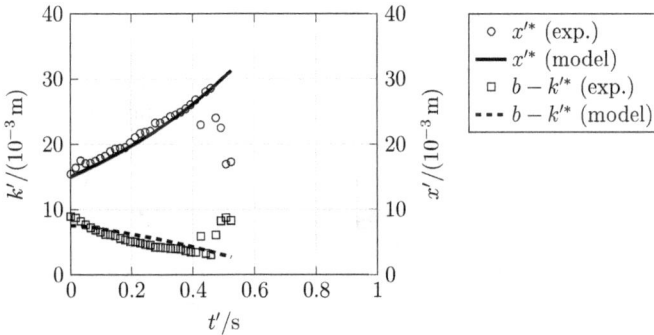

Figure 7.9: Time evolution of $x'^*$ and $k'^*$ during choking for experiment GR01034 with $l' = 15 \times 10^{-3}$ m and $\Delta Q'_{sc} = 2 \times 10^{-6}$ m$^3$ s$^{-1}$. Comparison of supercritical model and experiment results. Outliers may occur due to difficulty of postprocessing the moving interface.

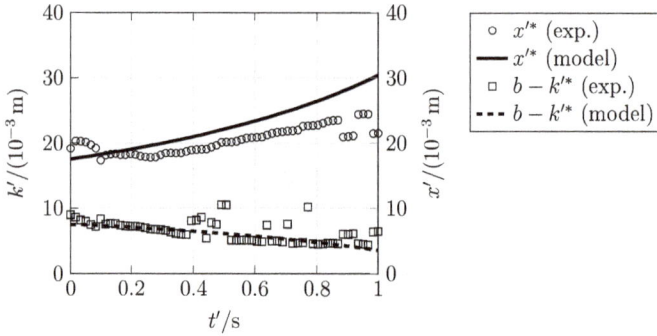

Figure 7.10: Time evolution of $x'^*$ and $k'^*$ during choking for experiment GR01051 with $l' = 20 \times 10^{-3}$ m and $\Delta Q'_{sc} = 0.5 \times 10^{-6}$ m$^3$ s$^{-1}$. Comparison of supercritical model and experiment results. Outliers may occur due to difficulty of postprocessing the moving interface.

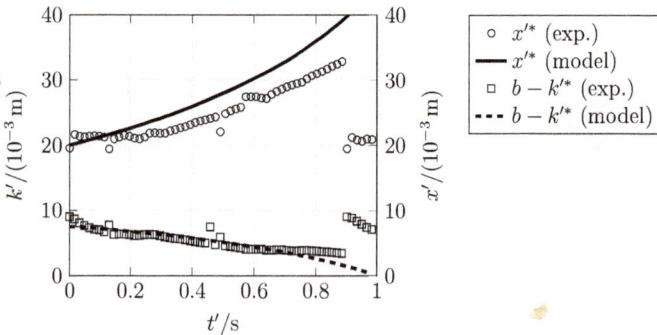

Figure 7.11: Time evolution of $x'^*$ and $k'^*$ during choking for experiment GR01053 with $l' = 20 \times 10^{-3}$ m and $\Delta Q'_{sc} = 1 \times 10^{-6}$ m$^3$ s$^{-1}$. Comparison of supercritical model and experiment results. Outliers may occur due to difficulty of postprocessing the moving interface.

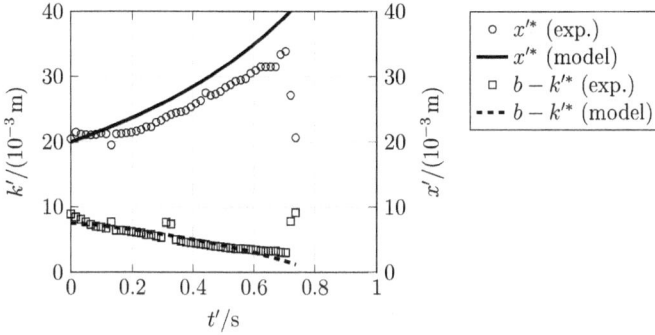

Figure 7.12: Time evolution of $x'^*$ and $k'^*$ during choking for experiment GR01055 with $l' = 10 \times 10^{-3}$ m and $\Delta Q'_{sc} = 1.5 \times 10^{-6}$ m$^3$ s$^{-1}$. Comparison of supercritical model and experiment results. Outliers may occur due to difficulty of postprocessing the moving interface.

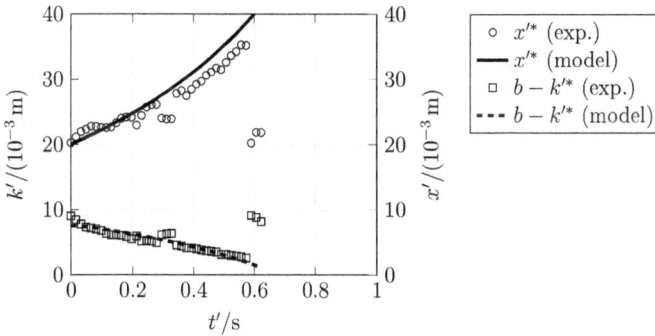

Figure 7.13: Time evolution of $x'^*$ and $k'^*$ during choking for experiment GR01059 with $l' = 20 \times 10^{-3}$ m and $\Delta Q'_{sc} = 2 \times 10^{-6}$ m$^3$ s$^{-1}$. Comparison of supercritical model and experiment results. Outliers may occur due to difficulty of postprocessing the moving interface.

# Chapter 8

# Summary and Outlook

This work presents a theoretical, numerical, and experimental analysis of forced, isothermal open capillary channel flow in a microgravity environment for single-phase and for two-phase bubbly flows. Prime focus of the analysis is the quantitative determination of the flow rate limitation in such flows and the behaviour of the gas-liquid interface for subcritical and supercritical flow rates. The shape and, ultimately, the stability of the interface is bound to the pressure development in the liquid flowing through the channel and the corresponding pressure difference between liquid and ambient gas in accordance with the established Young-Laplace equation (compare equation (2.3)). An extensive experimental study was undertaken in the unique environmental conditions onboard the ISS and the results thereof are used to validate existing and extended models used in numerical simulations. Additionally, the interface behaviour in the supercritical regime is investigated experimentally and a model is developed to describe the transient phenomenon of bubble ingestion based. This work is intended to contribute to the principle understanding of open capillary flows and to assist the design and development of fluid management systems in Space that exploit capillary forces for handling flowing liquids. Accurate and validated models are essential for the latter due to the limited possibilities for experimental design studies.

A one-dimensional model is developed for single-phase open capillary channel flow in a wedge-shaped channel with a triangular cross-section based on the work of Klatte [33], Rosendahl et al. [46], and Shapiro [53]. The single-phase model is then extended to allow for two-phase bubbly flow using a homogeneous flow model. Viscous pressure loss is modelled in identical manner for both single-phase and two-phase flow. Appropriate boundary conditions are postulated based on a constant contact static contact angle $\gamma = 0°$ and CFD simulations provide pressure boundary conditions for the range of investigated flow parameters. The derivation of the one-dimensional model is laid out in dimensional form; subsequent scaling reveals the underlying dimensionless numbers for this particular flow problem and channel geometry as Oh, $\tilde{l}$, and $\Lambda$. The one-dimensional model is solved using numerical methods outlined in chapter 5 to determine critical flow rates that represent the flow rate limitation in the presented open capillary channel. Critical flow rates and representative interface contours are presented for the parameter range investigated in the experiments on the ISS and compared with the acquired experiment data. The results illuminate the dependency of $Q_{crit}$ on $\tilde{l}$ and two distinct regimes are identified in which either the convective or viscous pressure terms are dominant for the pressure development in the channel. For this particular channel geometry and set of parameters, the critical flow rates determined with the one-dimensional model may be fitted to $Q_{crit} \approx 0.03\,\tilde{l}^{-0.5}$ for inertia dominated flows ($\tilde{l} < 10^{-3}$) and to $Q_{crit} \approx 0.015\,\tilde{l}^{-1}$ for viscous dominated flows ($\tilde{l} > 10^{-1}$) with a transition region inbetween.

Three-dimensional simulations are performed with the same flow parameters using the numerical solver interFoam that is included in the open source CFD tool OpenFOAM. Additional results are derived from the three-dimensional computational solutions including velocity profiles, and streamlines showing that three-dimensional effects such as flow development and flow separation are present which affect the accuracy of the one-dimensional model especially for $\tilde{l} < 10^{-3}$ because the additional pressure loss is not accounted for. While the presented CFD simulations show good agreement with the experiment data the model assumptions of the one-dimensional model contribute to deviating results for decreasing $\tilde{l}$ which are attributed to the neglect of additional pressure loss sources identified above.

An additional comparison and validation of the numerical methods is performed by comparing representative contours of the free surface. Contours represent the shape of the interface in the $xz$-plane and are displayed for channel lengths $l' = \{10, 15, 20, 25, 30\} \times 10^{-3}\,\text{m}$ and flow rates $Q' = \{0.8, 0.9, 1\} \times Q'_{crit}$. The comparison reveals that the numerical methods agree well for subcritical flow rates, but the contour of the interface at $Q'_{crit}$ display similar trends but deviations are more pronounced than for subcritical flow. This is attributed to the sensitivity of the contour to flow rate differences in the vicinity of $Q'_{crit}$ as shown in figures 5.4 and 5.5. In addition, the one-dimensional model does not account for various three-dimensional pressure loss sources that have been identified in the CFD simulations as already mentioned above.

Critical flow rates for two-phase bubbly flow are determined using the extended one-dimensional model for flow parameters observed in the experiments. The experiment data displays an enhancement of the critical flow rate that depends on the average gas flow rate injected into the test channel. The results of the one-dimensional model display the same behaviour and the two data sets are in adequate agreement when the absolute values of critical flow rate enhancement are compared considering the simplicity of the used model and the accuracy of the experiment setup.

An experimental study of the supercritical flow regime is presented, in which bubbles are ingested into the flow path due to the periodic collapse of the interface. The experiment results are presented for the test channels with rectangular (GR) and triangular (WE) cross-section and they illuminate the dependency of the ingestion frequency on the flow rate imposed by the pump while the volume of the ingested bubbles appears to remain relatively constant. This implication of a direct relationship between the demanded flow rate and the gas flow rate caused by bubble ingestion is investigated by developing a new model for the supercritical flow regime. The model is based on the assumption that local velocities in the flowing liquid must not exceed the capillary wave speed $v'_{ca}$. For supercritical flow (i.e. when the pump is set to demand a local velocity in excess of $v'_{ca}$), the flexible interface moves in flow direction at a velocity that depends on the demanded flow rate and the smallest cross-sectional area in the test channel. The model is developed for the test channel with a rectangular cross-section used in EU1 and the predicted time-dependent locations of $x'^*$ and $k'^*$ coincide with experiment results.

While the experiment data shown here validates the models used to simulate the presented flow model, additional work may be required to improve the one-dimensional model. Especially the influence of flow development in the entrance region upstream of the channel inlet and flow separation appear to limit the accuracy of the model. While models for flow development and additional pressure loss in the entrance region are available for channels with triangular cross-section [51], in this particular case the obstruction of the flow path in the entrance region complicates the matter significantly. However, for the channels used in EU1 with a rectangular cross-section (these channels did not contain bubble injection devices), the addition of entrance region effects contributes to the accuracy of the model as shown by [9] and [19]. Furthermore, [43] describe a method to include flow separation in a one-dimensional streamtube model. Their proposed method prevents recuperation of convective pressure loss downstream of $x'^*$. A similar modification of the model presented here may increase the accuracy of $Q'_{crit,1D}$ for $\tilde{l} < 0.05$.

The satisfactory accuracy of the numerical results for isothermal flow spark interest for the influence of temperature variability on the interface stability in open capillary channel flow. In the majority non-laboratory environments, the assumption of isothermal flow is unlikely to hold true. Even large local temperature variations may be of interest in heat pipe applications or bioreactors. While the physical properties of the flowing liquid are most likely to affect the pressure loss within the channel, additional non-isothermal effects may include surface tension variations and non-constant contact angles. Additional research is therefore required to determine how temperature-dependencies may affect interface stability in the presented flow problem.

# Appendix A

# Indicial Notation for Tensor Operations

Comparison of vector notation and index notation.

|         | vector notation | index notation |
|---------|-----------------|----------------|
| scalar  | $a$             | $a$            |
| vector  | $\vec{a}$       | $a_i$          |
| tensor  | $\underline{\underline{A}}$ | $a_{ij}$ |

# Gradient

$$\mathrm{grad}(a) = \nabla a = \frac{\partial a}{\partial x_i} = \begin{bmatrix} \dfrac{\partial a}{\partial x_1} \\[2ex] \dfrac{\partial a}{\partial x_2} \\[2ex] \dfrac{\partial a}{\partial x_3} \end{bmatrix}$$

$$\mathrm{grad}(\vec{a}) = \nabla \vec{a} = \frac{\partial a_i}{\partial x_j} = \begin{bmatrix} \dfrac{\partial a_1}{\partial x_1} & \dfrac{\partial a_1}{\partial x_2} & \dfrac{\partial a_1}{\partial x_3} \\[2ex] \dfrac{\partial a_2}{\partial x_1} & \dfrac{\partial a_2}{\partial x_2} & \dfrac{\partial a_2}{\partial x_3} \\[2ex] \dfrac{\partial a_3}{\partial x_1} & \dfrac{\partial a_3}{\partial x_2} & \dfrac{\partial a_3}{\partial x_3} \end{bmatrix}$$

# Divergence

$$\mathrm{div}(\vec{a}) = \nabla \cdot \vec{a} = \frac{\partial a_i}{\partial x_i} = \frac{\partial a_1}{\partial x_1} + \frac{\partial a_2}{\partial x_2} + \frac{\partial a_3}{\partial x_3}$$

$$\text{div}(\underline{A}) = \nabla \cdot \underline{\underline{A}} = \frac{\partial a_i}{\partial x_j} = \begin{bmatrix} \dfrac{\partial a_{11}}{\partial x_1} + \dfrac{\partial a_{12}}{\partial x_2} + \dfrac{\partial a_{13}}{\partial x_3} \\[2mm] \dfrac{\partial a_{21}}{\partial x_1} + \dfrac{\partial a_{22}}{\partial x_2} + \dfrac{\partial a_{23}}{\partial x_3} \\[2mm] \dfrac{\partial a_{31}}{\partial x_1} + \dfrac{\partial a_{32}}{\partial x_2} + \dfrac{\partial a_{33}}{\partial x_3} \end{bmatrix}$$

# Rotation

$$\text{rot}(\vec{a}) = \nabla \times \vec{a} = -\varepsilon_{ijk}\frac{\partial a_i}{\partial x_j} = \begin{bmatrix} \dfrac{\partial a_3}{\partial x_2} - \dfrac{\partial a_2}{\partial x_3} \\[2mm] \dfrac{\partial a_1}{\partial x_3} - \dfrac{\partial a_3}{\partial x_1} \\[2mm] \dfrac{\partial a_2}{\partial x_1} - \dfrac{\partial a_1}{\partial x_2} \end{bmatrix}$$

$$\varepsilon_{ijk} = \begin{cases} 1 & \text{if } ijk = 123, 231, \text{ or } 312 \\ -1 & \text{if } ijk = 132, 213, \text{ or } 321 \\ 0 & \text{when } i = j = k \end{cases}$$

# Laplace Operator

$$\Delta a = \text{div}(\text{grad}(a)) = \nabla \cdot (\nabla a) = \nabla^2 a = \frac{\partial^2 a}{\partial x_j^2} = \frac{\partial^2 a}{\partial x_1^2} + \frac{\partial^2 a}{\partial x_2^2} + \frac{\partial^2 a}{\partial x_3^2}$$

$$\Delta \vec{a} = \text{div}(\text{grad}(\vec{a})) = \nabla \cdot (\nabla \vec{a}) = \nabla^2 \vec{a} = \frac{\partial^2 a_i}{\partial x_j^2} = \begin{bmatrix} \dfrac{\partial^2 a_1}{\partial x_1^2} + \dfrac{\partial^2 a_1}{\partial x_2^2} + \dfrac{\partial^2 a_1}{\partial x_3^2} \\[2mm] \dfrac{\partial^2 a_2}{\partial x_1^2} + \dfrac{\partial^2 a_2}{\partial x_2^2} + \dfrac{\partial^2 a_2}{\partial x_3^2} \\[2mm] \dfrac{\partial^2 a_3}{\partial x_1^2} + \dfrac{\partial^2 a_3}{\partial x_2^2} + \dfrac{\partial^2 a_3}{\partial x_3^2} \end{bmatrix}$$

# Appendix B

# CCF EGSE Flow Map

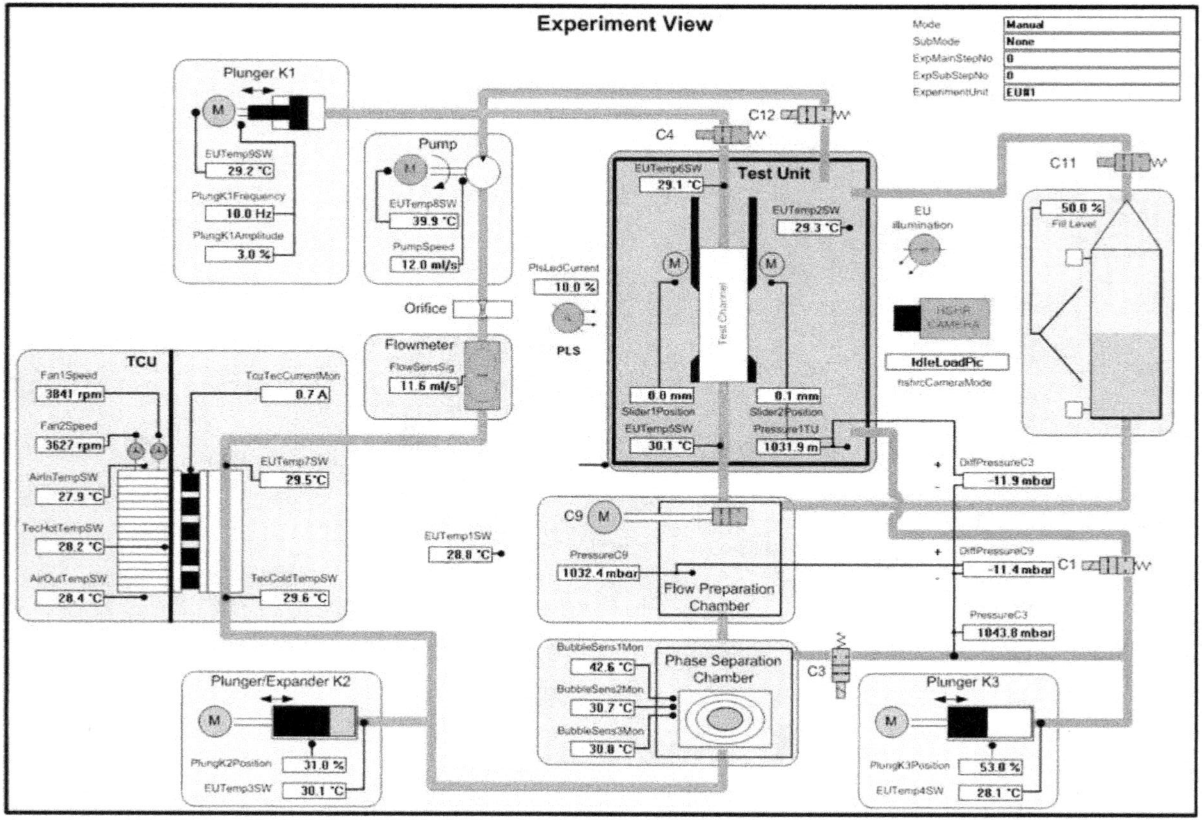

Figure B.1: Map of the fluid loop as displayed in the control software for the CCF experiment displaying relative locations of various important components.

# Appendix C

# Tabulated Ayyaswammy Velocity

Tabulated results for $\bar{v}_{Ayy}$ as a function of the effective contact angle $\gamma_{eff}$ (rows) and the wedge half angle $\alpha$ (columns) [2].

| | 1° | 5° | 10° | 20° | 30° | 40° | 50° | 60° |
|---|---|---|---|---|---|---|---|---|
| 0.1° | 4.8578E-05 | 1.0156E-03 | 3.2329E-03 | 8.0790E-03 | 1.1092E-02 | 1.1568E-02 | 9.9138E-03 | 6.9893E-03 |
| 0.5° | 4.8590E-05 | 1.0169E-03 | 3.2409E-03 | 8.1204E-03 | 1.1183E-02 | 1.1707E-02 | 1.0084E-02 | 7.1688E-03 |
| 1° | 4.8605E-05 | 1.0184E-03 | 3.2509E-03 | 8.1723E-03 | 1.1297E-02 | 1.1881E-02 | 1.0299E-02 | 7.3929E-03 |
| 5° | 4.8720E-05 | 1.0306E-03 | 3.3333E-03 | 8.5885E-03 | 1.2223E-02 | 1.3320E-02 | 1.2112E-02 | 9.3135E-03 |
| 10° | 4.8859E-05 | 1.0454E-03 | 3.4270E-03 | 9.1117E-03 | 1.3419E-02 | 1.5238E-02 | 1.4610E-02 | 1.2095E-02 |
| 15° | 4.8992E-05 | 1.0598E-03 | 3.5220E-03 | 9.6367E-03 | 1.4650E-02 | 1.7266E-02 | 1.7347E-02 | 1.5294E-02 |
| 20° | 4.9120E-05 | 1.0736E-03 | 3.6151E-03 | 1.0162E-02 | 1.5910E-02 | 1.9398E-02 | 2.0303E-02 | 1.8872E-02 |
| 25° | 4.9244E-05 | 1.0871E-03 | 3.7063E-03 | 1.0687E-02 | 1.7197E-02 | 2.1622E-02 | 2.3466E-02 | 2.2813E-02 |
| 30° | 4.9363E-05 | 1.1002E-03 | 3.7959E-03 | 1.1213E-02 | 1.8509E-02 | 2.3936E-02 | 2.6831E-02 | 2.7117E-02 |
| 35° | 4.9479E-05 | 1.1131E-03 | 3.8842E-03 | 1.1739E-02 | 1.9848E-02 | 2.6343E-02 | 3.0404E-02 | 2.7117E-02 |
| 40° | 4.9592E-05 | 1.1257E-03 | 3.9716E-03 | 1.2269E-02 | 2.1217E-02 | 2.8847E-02 | 3.4194E-02 | 2.7117E-02 |
| 45° | 4.9704E-05 | 1.1382E-03 | 4.0586E-03 | 1.2804E-02 | 2.2621E-02 | 3.1460E-02 | 3.4194E-02 | 2.7117E-02 |
| 50° | 4.9814E-05 | 1.1505E-03 | 4.1456E-03 | 1.3347E-02 | 2.4067E-02 | 3.4194E-02 | 3.4194E-02 | 2.7117E-02 |
| 55° | 4.9923E-05 | 1.1629E-03 | 4.2331E-03 | 1.3900E-02 | 2.5563E-02 | 3.4194E-02 | 3.4194E-02 | 2.7117E-02 |
| 60° | 5.0033E-05 | 1.1754E-03 | 4.3216E-03 | 1.4467E-02 | 2.7117E-02 | 3.4194E-02 | 3.4194E-02 | 2.7117E-02 |
| 65° | 5.0143E-05 | 1.1880E-03 | 4.4118E-03 | 1.5051E-02 | 2.7117E-02 | 3.4194E-02 | 3.4194E-02 | 2.7117E-02 |
| 70° | 5.0256E-05 | 1.2008E-03 | 4.5042E-03 | 1.5657E-02 | 2.7117E-02 | 3.4194E-02 | 3.4194E-02 | 2.7117E-02 |
| 75° | 5.0371E-05 | 1.2140E-03 | 4.5995E-03 | 1.5657E-02 | 2.7117E-02 | 3.4194E-02 | 3.4194E-02 | 2.7117E-02 |
| 80° | 5.0490E-05 | 1.2276E-03 | 4.6985E-03 | 1.5657E-02 | 2.7117E-02 | 3.4194E-02 | 3.4194E-02 | 2.7117E-02 |
| 85° | 5.0613E-05 | 1.2480E-03 | 4.6985E-03 | 1.5657E-02 | 2.7117E-02 | 3.4194E-02 | 3.4194E-02 | 2.7117E-02 |
| 89° | 5.0715E-05 | 1.2480E-03 | 4.6985E-03 | 1.5657E-02 | 2.7117E-02 | 3.4194E-02 | 3.4194E-02 | 2.7117E-02 |

# Appendix D

# Numerical Study - Pipe flow (simpleFoam)

## Introduction

Comparing results of CFD simulations with extensively discussed and well known analytical solutions reinforce the validity of the software's code for comparable cases for which an analytical solution is not available. In this case, a valid numerical algorithm is required to simulate three-dimensional steady laminar flow through a closed duct. To this end, the open source CFD tool package OpenFOAM v2.1.1 is utilised to simulate steady incompressible laminar flow through a cylindrical duct for multiple Reynolds numbers in the rage of $100 < \text{Re} < 1500$. The results of the simulations are then compared to analytical solutions of flow velocity, pressure loss, and flow behaviour within the entrance region of the duct. OpenFOAM uses finite volume methods and natively provides a solver algorithm called simpleFoam , which is a steady-state solver for incompressible turbulent flow, but can also be modified to compute laminar flow by excluding the turbulence models in the simulation.

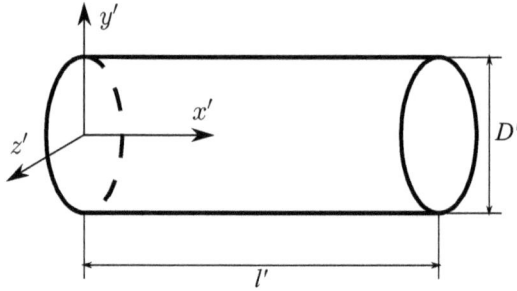

Figure D.1: Geometry of the cylindrical duct discussed in this section. The duct has a diameter $D'$ and a length $l'$. Flow is parallel to the $x'$-axis and the origin of coordinates is located at the inlet of the channel.

## Analytical solution

An analytical solution for the present problem is required to evaluate the accuracy of the numerical solution.

## Fully developed flow

Steady incompressible duct flows are laminar within the Reynolds number range $Re_{crit} < 2300$. Above this critical Reynolds number flow undergoes a transition to turbulent flow. The cases that are considered in this section are all below $Re = 2300$ and therefore laminar flow can be assumed for the analysis. Fully developed laminar flow in a cylindrical duct can be solved analytically with regard to the pressure gradient in flow direction and the flow field.

Consider a cylindrical duct with a length $l'$, a constant diameter $D'$, and a constant cross-sectional area $A' = \pi D'^2/4$ (see figure D.1). Flow is fully developed at the inlet and thus the only non-zero velocity component is in the direction of flow which is in the direction of the $x'$-axis. The force balance equation for a fluid element within the duct is composed of a pressure force in flow direction and a friction force at the wall and can be written as

$$p'(x')\pi r'^2 - p'(x' + \mathrm{d}x')\pi r'^2 = \tau'(r')2\pi r'\mathrm{d}x' \quad , \tag{D.1}$$

with pressure $p'$ at locations $x'$ and $x' + \mathrm{d}x'$ and shear stress $\tau'$ as a function of the radius $r'$. Assuming a newtonian fluid, shear stress in the fluid can be described by

$$\tau(r') = \mu\frac{\mathrm{d}v'}{\mathrm{d}r'} \quad , \tag{D.2}$$

with the velocity in flow direction $v'$ and the dynamic viscosity of the fluid $\mu$. Combining equations (D.1) and (D.2) leads to the differential equation

$$\frac{\mathrm{d}v'}{\mathrm{d}r'} = \frac{\mathrm{d}p'}{\mathrm{d}x'}\frac{r'}{2\mu} \tag{D.3}$$

which in turn is integrated and solved using a zero velocity boundary condition at the wall $v'(r' = R') = 0$ to reveal a parabolic velocity profile as a function of $r'$

$$v'(r') = -\frac{\mathrm{d}p'}{\mathrm{d}x'}\frac{R'^2}{4\mu}\left[1 - \left(\frac{r'}{R'}\right)^2\right] \quad . \tag{D.4}$$

The maximum velocity is located at $r' = 0$ and is

$$v'_{max} = v'(r' = 0) = -\frac{\mathrm{d}p'}{\mathrm{d}x'}\frac{R'^2}{4\mu} \quad . \tag{D.5}$$

Integrating the velocity over the cross-sectional area reveals the volumetric flow rate $Q'$

$$Q' = \int_{A'} v'(r')\mathrm{d}A' = -\frac{\mathrm{d}p'}{\mathrm{d}x'}\frac{\pi R'^4}{8\mu} \quad . \tag{D.6}$$

And with $Q' = \bar{v}'A'$, the velocity averaged over any cross-section is found to be

$$\bar{v}' = -\frac{\mathrm{d}p'}{\mathrm{d}x'}\frac{R'^2}{8\mu} = \frac{1}{2}v'_{max} \quad . \tag{D.7}$$

Assuming a linear pressure loss in flow direction, finding the pressure loss between two points in the duct that are a distance $l'$ along the $x'$-axis from each other is then simply a matter of exchanging $-\mathrm{d}p'/\mathrm{d}x'$ for $\Delta p'/l'$ in equation (D.7)

$$\frac{\Delta p'}{l'} = \frac{32\mu\bar{v}'}{D'^2} \quad , \tag{D.8}$$

and rearranging for $\Delta p'$ and expanding by $\rho\bar{v}'$ yields

$$\Delta p' = \frac{\rho\bar{v}'^2}{2}\frac{64}{\mathrm{Re}}\frac{l'}{D'} \quad , \tag{D.9}$$

with the Reynolds number $\text{Re} = \rho \bar{v}' D' / \mu$. The non-dimensional pressure loss $\Delta p$ is found by scaling with the characteristic pressure $p_c = (\rho \bar{v}'^2)/2$ ,

$$\Delta p = \frac{\Delta p'}{p_c} = K_f \frac{l'}{D'} \quad ; \quad K_f = \frac{64}{\text{Re}} \quad , \tag{D.10}$$

and we obtain the friction factor $K_f$. Note that this particular linear relation of non-dimensional pressure loss to Re only applies for fully developed laminar flow in a cylindrical tube.

## Entrance region

The entrance length $l'_e$ of a duct is defined as the length within which flow develops from some initial velocity profile at the inlet to a fully developed profile further downstream. In this region the pressure loss is not equal to that of the fully developed region due to pressure gradients and non-zero velocities normal to the axis of the cylindrical duct. Different models have been proposed to determine the length of the entrance region analytically and usually follow the assumption that flow develops from a homogeneous velocity profile with $\partial v' / \partial y' = \partial v' / \partial z' = 0$ at $x' = 0$. Shah and London [52] define the entrance length somewhat arbitrarily as the distance required for the centreline velocity to reach $0.99\, v'_{max}$. The entrance length is then found to be

$$\frac{l'_e}{D'} \approx 0.06\text{Re} \tag{D.11}$$

Within the entrance length the boundary layer grows due to shear stress until it converges at the axis of the cylindrical duct, thus generating the parabolic velocity profile derived above in equation (D.4). The growth of the boundary layer accounts for an additional pressure loss that is non-linear in nature and decreases with increasing $\hat{x}$. Various models have been formulated to compute the pressure drop in the entrance region and a summary is given by Stange in [59]. One analysis of the entrance region is performed by Sparrow and Lin [56]. In their work, they develop a method to analyse the additional pressure drop within an entrance region of a duct independent of its cross-section and apply it specifically to flow within a circular tube and a parallel-plate channel. In their findings they state that the method is 'fully adequate for any practical purpose' and validate it by comparing the model with pressure drop experiments in the entrance region that were performed by Shapiro et al. [54]. In accordance with this model the pressure loss in the entrance region is defined in non-dimensional form as

$$\Delta p = K_f \frac{x'}{D'} + K(x') \tag{D.12}$$

where $K_f$ is the Darcy friction factor for fully developed flow, which has already been determined, and $K(x')$ is the additional pressure loss due to the development of the boundary layer as a function of $x'$, which in the case of the circular tube is described by

$$K(x') = \frac{4}{3} + \sum_{i=1}^{\infty} \frac{8}{\alpha_i^2} \left( e^{-4\alpha_i^2 \frac{x'}{D'\text{Re}}} - 3 \right) e^{-4\alpha_i^2 \frac{x'}{D'\text{Re}}} \tag{D.13}$$

where $\alpha_i$ are eigenvalues of which the first 25 are listed in [56] and are used here to compute $K(x')$ to predict the additional pressure loss in the entrance region.

# Numerical solution

Numerical simulations were performed using the CFD toolbox OpenFOAM for Re $\in$ $\{100, 200, 300, 400, 500, 600, 700, 800, 900, 1000, 1250, 1500\}$.

## Case setup

A mesh is generated for the numerical simulations based on the geometry displayed in figure (D.1). Considering that this particular problem should depend solely on the Reynolds number, arbitrary values for geometrical dimensions and fluid properties are chosen to simplify the execution of the simulations. The cylinder geometry is described by its diameter $D' = 1$ m and its length $l' = 1.5 l'_e$. The entrance length is assumed to be $l'_e = 0.06 D' \text{Re}$. The dynamic viscosity of the fluid is set to 0.01 m$^2$/s. Thus the Reynolds number is defined by the mean velocity so that Re $= 100 \bar{v}'$.

The mesh is generated in blockMesh, the native meshing utility of OpenFOAM, and it consists entirely of hexahedral cells (see figure D.2). A base cell length of $\delta x' = 0.03762$ m is used to discretise the geometry spatially in the direction of flow. The discretisation of the cross-section of the cylinder is identical for each Reynolds number and is graded towards the wall. The number of cells in the mesh depends linearly on the entrance length with $N = 528 \, l'/\delta x'$. The boundary conditions for the simulation are listed in table D.1.

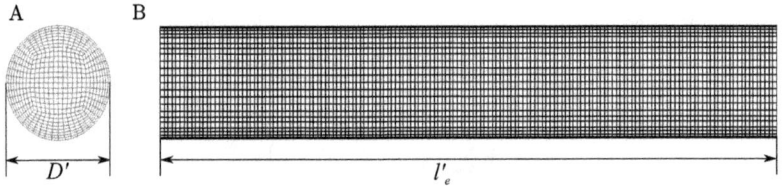

Figure D.2: The mesh for the geometry of the cylindrical pipe is generated using an OpenFOAM utility. The cells in the cross-section (A) are graded towards the wall. The length of the mesh is set to $l' = 1.5 \, l'_e$, only the entrance region is displayed here (B). The mesh consists entirely of hexahedral cells.

Table D.1: Boundary conditions for the numerical simulation.

|  | inlet | outlet | walls |
|---|---|---|---|
| patch type | patch | patch | wall (no slip) |
| $v'$ in $\frac{\mathrm{m}}{\mathrm{s}}$ | $0.01\mathrm{Re}$ | $\nabla v'_i = 0$ | $0$ |
| $p'$ in $\frac{\mathrm{kg}}{\mathrm{ms}^2}$ | $\nabla p' = 0$ | $0$ | $\nabla p' = 0$ |

The governing equations of this flow problem are the continuity equation for steady incompressible flow

$$\nabla \cdot v'_i = 0 \tag{D.14}$$

and the momentum equation

$$(v'_j \cdot \nabla)v'_i = -\nabla p' + \mu \nabla^2 v'_i \tag{D.15}$$

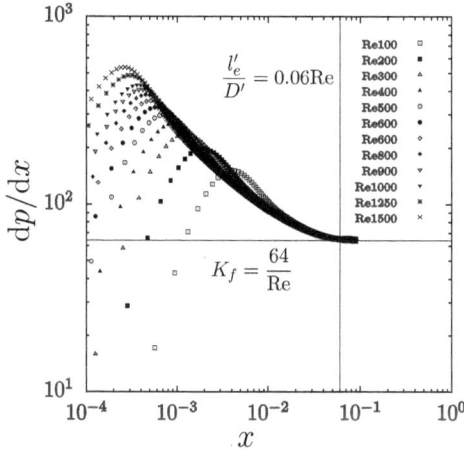

Figure D.3: Results for the scaled pressure gradient along the centreline of the cylindrical pipe. The assumed scaled length of the entrance region is plotted as well as the theoretical pressure gradient for fully developed flow. The pressure gradient tends towards the theoretical value across the entire range of simulations.

The OpenFOAM solver simpleFoam is used to the simulate steady incompressible flow through the cylindrical pipe. This solver uses the SIMPLE algorithm, which is an implicit pressure correction method for incompressible flow. Though the solution algorithm of simpleFoam contains a turbulence model, it can be switched off to simulate laminar flow. The numerical schemes that are used in the simulations are upwind for divergence terms, linear (central differences) for gradient terms, and Gaussian with linear interpolation for Laplacian terms.

## Results and discussion

In order to compare the results of the various simulations the following scaling method was applied

$$x = \frac{x'}{D'\mathrm{Re}} \quad ; \quad p = \frac{2p'}{\rho \bar{v}'^2} \quad ; \quad v = \frac{v'}{\bar{v}'} \quad ; \quad \mathrm{Re} = \frac{\rho \bar{v}' D'}{\mu} \tag{D.16}$$

Postprocessing of the numerical simulations is generally performed along the centreline of the cylindrical pipe. The pressure gradient along the centreline is displayed in figure D.3 for each of the observed Reynolds numbers. The entrance length and theoretical pressure gradient for fully developed laminar flow are also plotted in the figure. The results show that the numerical solution tends towards the analytical solution at the end of the entrance length. The difference between the numerical simulation and the analytical solution is $< 3\%$ at $x' = l'_e$. However, one should note that 0.06Re is only an approximation for $l'_e/D'$ and flow is not quite fully developed in the numerical simulation as can be seen in figure D.5.

Within the entrance region the pressure gradient rises before decreasing again towards $K_f = 64/\text{Re}$. This is unexpected since entrance effects should increase the pressure gradient near the inlet and subside while the boundary layer grows until the additional pressure loss equals zero once the flow is fully developed. The unexpected behaviour of the plotted results is due to the method of post-processing. In the entrance region, there is a pressure gradient over the cross-section because the boundary layer is evolving. Figure D.3 however only uses data from the centreline.

Averaging pressure over the cross-section yields a more expected result, as can be seen in figure D.4. Here the predictions for the pressure development along the pipe are displayed in blue. The predictions are based on the analytical solution for fully developed flow and the model by Sparrow and Lin [56]. The results of the numerical simulations are presented in red with dots representing values extracted from the centreline and crosses representing values for the cross-sectional area averaged pressure $\bar{p}'$. Additionally, experimental values from Shapiro et al. [54] are plotted as a regression in magenta. The results for $\bar{p}'$ coincide very well with the expected values over the entire length of the channel.

In figure (D.5) the velocity at $(x', y', z') = (l'_e, 0, 0)$ is compared between the simulation and $v_{max}$ for fully developed flow. The results of the numerical simulation are in good agreement with the analytical solution. At the end of the assumed entrance region the difference between the theoretical value of $v_{max} = 2\bar{v}$ and the

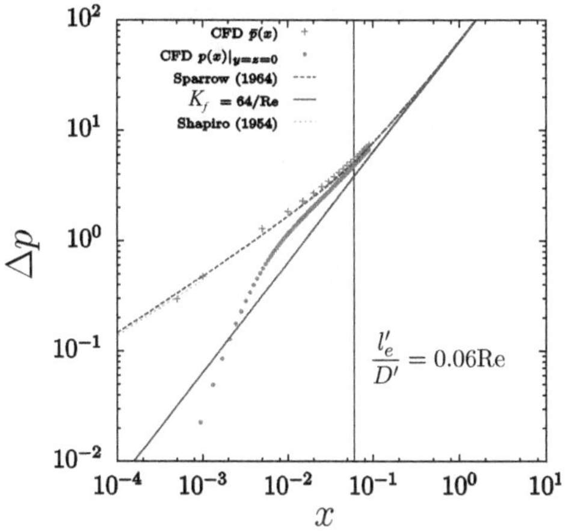

Figure D.4: Comparison of the numerical data with the model by Sparrow and Lin [56] for the entrance region. The numerical results are averaged over the cross-section (--) and also displayed as centerline data (•). Results from [54] are displayed in magenta. The results of the numerical simulations agree well with the predictions over the entire length of the channel when $\bar{p}'$ is used.

results of the simulations are $< 2\,\%$. The velocity profile at $x' = l'_e$ is shown in figure (D.6). The velocity profile in flow direction is parabolic as it is expected to be based on the analytical solution.

The good agreement between the numerical simulations and the analytical solution shows that OpenFOAM's solver simpleFoam is a valid choice for determining pressure loss in steady laminar incompressible flow in closed ducts. That being said, the influence of the mesh refinement on the numerical results must be determined. Also, one should note that a structured mesh is used that consists entirely of hexahedral cells. The numerical methods used in OpenFOAM are most accurate when applied to hexahedral cells. Use of an unstructured grid with tetrahedral cells may influence the results significantly.

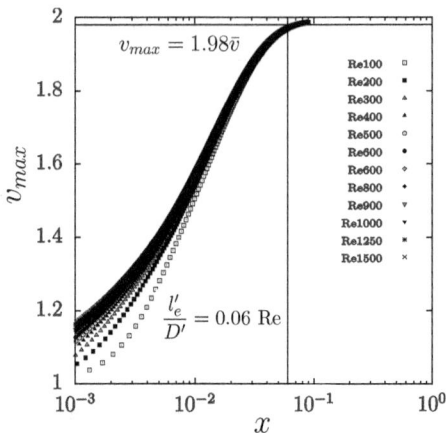

Figure D.5: Velocity in flow direction along the centreline for all of the simulations. The development of the boundary layer in the entrance region is visible as velocity increases along the centreline. At $x' = l'_e$ flow is almost fully developed. For practical purposes, flow is often assumed to be fully developed when $v_{max}$ reaches $1.98\bar{v}$.

## Mesh analysis

The influence of the spatial resolution of the mesh on the numerical results should tend towards zero for a mesh independent solution. Determining the influence of the mesh resolution on the numerical results is determined by refining the mesh resolution iteratively by a constant factor. Let the coarsest level of refinement have a cell length of $\delta x' = 0.078$ m. Each level of refinement is achieved by dividing the previous level's cell length by an arbitrary factor. In this case the cell length of a given level of refinement is defined as

$$\delta x'_n = 0.078 \cdot 1.2^{-n} \qquad (D.17)$$

where $n$ is the level of refinement and $n = 0$ is the coarsest level. The mesh analysis is performed for Re $= 100$ and $0 \leq n \leq 8$. The number of cells $N_n$ in the meshes used in this analysis increases with $n$ ranges from $N_0 = 1.5 \cdot 10^4$ to $N_8 = 1.3 \cdot 10^6$. The numerical results in the previous sections are attained using a mesh refinement level of $n = 4$.

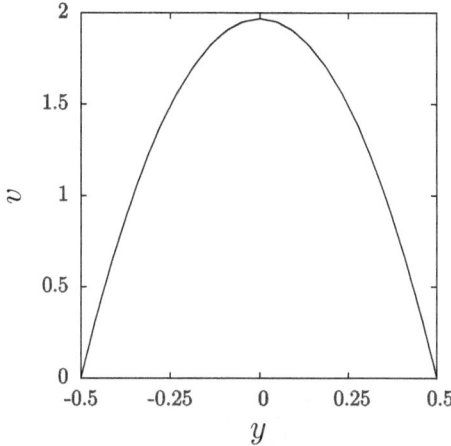

Figure D.6: Velocity profile along the $y$-axis at $x' = l'_e$. The velocity profile is parabolic as predicted by the analytical solution. The maximum value for $v$ is at the centreline and almost conforms to the analytical solution of $v_{max} = 2\bar{v}$.

Figures D.7 and D.8 show the results of centreline pressure gradient and velocity respectively for the various levels of mesh refinement. The results indicate that the difference to the analytical solution decreases with increasing $n$, but the effectiveness of mesh refinement is almost negligible for $n > 6$. Figures D.9a and D.9b reinforce this statement. At the assumed entrance length $l'_e = 0.06D'\text{Re}$ the numerical results for pressure gradient and velocity converge towards the analytical solution. Based on these findings, it can be assumed that the numerical results of pressure gradient and velocity along the centreline are independent of the mesh resolution for $n \geq 4$ with an error of $< 3\%$.

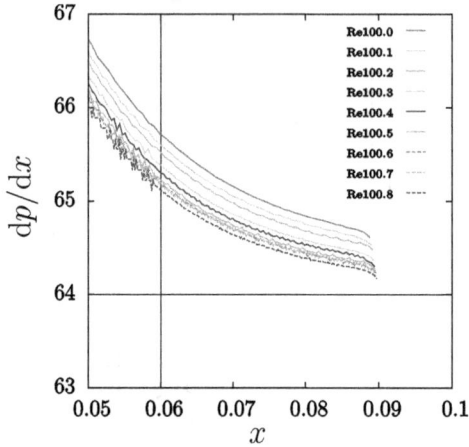

Figure D.7: Overview of $\mathrm{d}p/\mathrm{d}x$ for the various mesh refinement levels. Mesh refinement results in values closer to the theoretical value but the effectiveness of mesh refinement appears to diminish for $n > 6$.

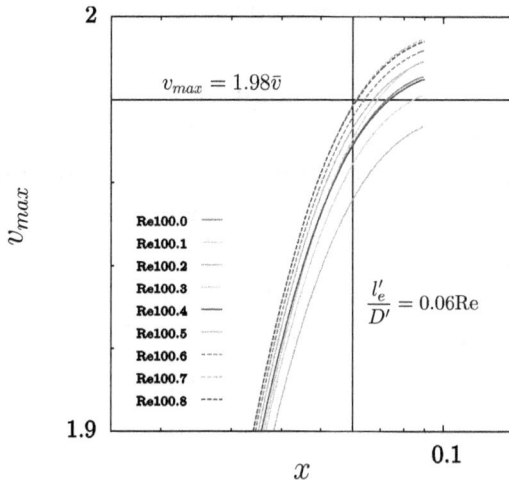

Figure D.8: Overview of $v_{max}$ at $x' = l'_e$ for the different levels of mesh refinement. Higher levels of refinement approach the analytical solution earlier. Refinement seems to lose effectiveness for $n > 6$.

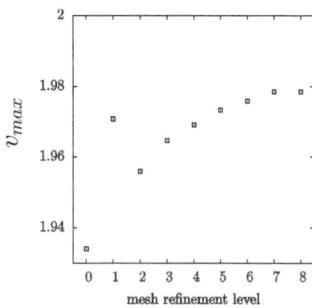

(a) Mesh analysis of the velocity at the centreline.

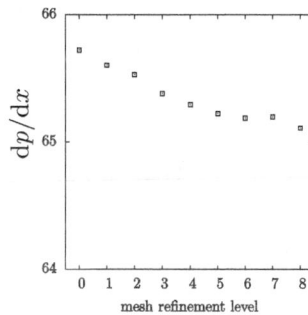

(b) Mesh analysis of the pressure gradient at the centreline.

Figure D.9: Overview of the effectiveness of mesh refinement based on the dimensionless centreline values of pressure gradient and velocity at $x' = l'_e$. The plots both display a convergence towards higher accuracy for higher levels of refinement.

# Appendix E

# Numerical Study - Capillary Rise (interFoam)

## Introduction

A basic test case is considered to determine whether the OpenFOAM solver interFoam can be used to simulate two-phase flow problems . The test case is two-dimensional capillary rise along a vertical wall with an equilibrium height $z_0'$ and a static contact angle of $\gamma = 0°$. An analytical solution is available to validate

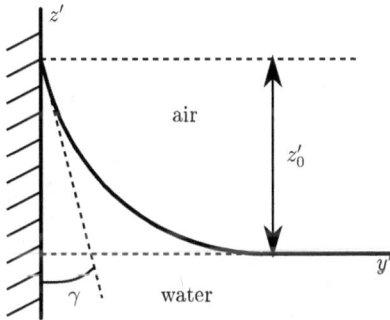

Figure E.1: Two-dimensional capillary rise at a vertical wall.

numerical results. The solution procedure is outlined in [21] and only the relevant solutions are presented here. The scenario is displayed in figure E.1. Far from the wall, $y \to \infty$, the liquid height $z'$ is Zero. The height of the meniscus at the wall can be analytically determined because hydrostatic and capillary forces are in equilibrium. According to the Young-Laplace equation, the capillary pressure $p'_{cap}$ is determined by the local curvature and the surface tension $\sigma$:

$$\Delta p'_{cap} = -\sigma \left( \frac{1}{R'_1} + \frac{1}{R'_2} \right). \tag{E.1}$$

In a two dimensional problem, $R'_2 = \infty$ and $R'_1$ is defined as:

$$\frac{1}{R'_1} = \frac{\dfrac{\partial^2 z'}{\partial y'^2}}{\left[ 1 + \left( \dfrac{\partial z'}{\partial y'} \right)^2 \right]^{3/2}}. \tag{E.2}$$

Hydrostatic pressure is defined as the difference between the phase pressures in the vicinity of the interface:

$$p'_g - p'_l = p'_a - \rho_G g' z' - p'_a + \rho_L g' z', \tag{E.3}$$

where $\rho_G$ and $\rho_L$ are the respective densities of gas and liquid and $g'$ is the gravitational acceleration. The balance equation for this problem is then defined as:

$$\frac{\Delta \rho g' z'}{\sigma} - \frac{\dfrac{\partial^2 z'}{\partial y'^2}}{\left[ 1 + \left( \dfrac{\partial z'}{\partial y'} \right)^2 \right]^{3/2}} = 0. \tag{E.4}$$

The height of the interface at the wall is calculated to be:

$$z'_0 = L'_c [2(1 - \sin \theta)]^{1/2}, \tag{E.5}$$

with the characteristic length $L'_c = \sqrt{\dfrac{\sigma}{\Delta \rho g'}}$ and the static contact angle $\theta$. Furthermore, the contour of the interface is described by the following equation:

$$y' = L'_c \left[ \operatorname{arcosh} \frac{2L'_c}{z'} - \operatorname{arcosh} \frac{2L'_c}{z'_0} + \left( 4 - \frac{z'^2}{L'^2_c} \right)^{1/2} - \left( 4 - \frac{z'^2_0}{L'^2_c} \right) \right], \tag{E.6}$$

atmosphere

wall

air

symmetry

water

inlet

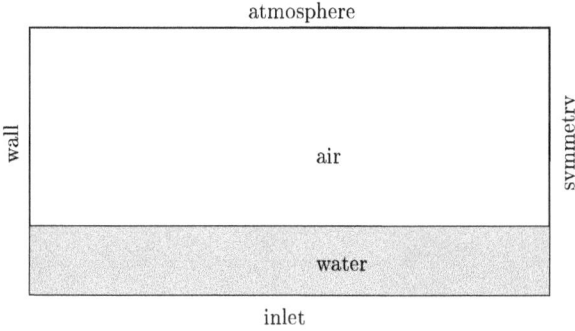

Figure E.2: Initial conditions of the presented test case.

# Numerical Setup

The computational domain of the CFD simulation is displayed in figure E.2. The geometry of the domain is defined as $\Delta z' = 20$ mm and $\Delta y' = 40$ mm. The chosen fluids are water ($\beta = 1$) and air ($\beta = 0$) at $T' = 20$ °C. The respective densities of the liquid and gas are $\rho_L = 998.21\,\mathrm{kg\,m^{-3}}$ and $\rho_G = 1.1885\,\mathrm{kg\,m^{-3}}$. The respective kinematic viscosities of the liquid and gas are $\nu_L = 1.003 \times 10^{-6}\,\mathrm{m^2\,s^{-1}}$ and $\nu_G = 15.32 \times 10^{-6}\,\mathrm{m^2\,s^{-1}}$. Gravitational acceleration is $g' = 9.81\,\mathrm{m\,s^{-2}}$ and the contact angle is $\gamma = 0°$. Surface tension is defined as $\sigma = 0.072\,74\,\mathrm{N\,m^{-1}}$.

A two-dimensional problem is defined in the VOF method used by interFOAM by setting the appropriate boundary condition ('empty') in the neglected dimension. A symmetry plane is applied to the right boundary. The left boundary is defined as a solid wall with a no-slip velocity boundary condition and a contact angle of $0°$. The upper boundary is defined to simulate an open gaseous atmosphere, the lower is a simple pressure based inlet to allow liquid to enter the domain as required. The domain is discretized into a mesh with an aspect ratio of 1:1:1 and cell lengths are universally $\delta x' = 0.2$ mm. In the vicinity of the interface, the mesh is refined to $\delta x' = 0.1$ mm.

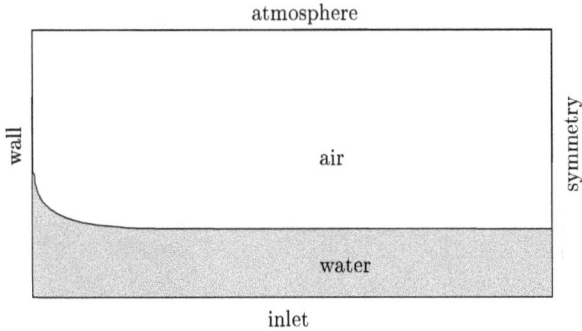

Figure E.3: Numerical solution of final equilibrium condition of the presented test case.

## Results and Discussion

The numerically determined equilibrium height of the capillary meniscus is displayed in figure E.3. The contour of the interface is plotted against the analytical solution in figure E.4. The equilibrium capillary height at the wall is $z'_{0,a} = 3.8567 \times 10^{-3}\,\text{m}$ for the analytical solution and $z'_{0,OF} = 4.029\,14 \times 10^{-3}\,\text{m}$. The absolute error of the numerical solution may be defined as:

$$\varepsilon_{OF} = \frac{|z'_{0,a} - z'_{0,OF}|}{z'_{0,a}} = 0.047 \,\hat{=}\, 0.173 \times 10^{-3}\,\text{m}. \tag{E.7}$$

While an absolute error of less than 5 % to the analytical solution may at first seem quite high, one must also consider that the mesh resolution in the vicinity of the interface is $\delta x' = 0.1 \times 10^{-3}\,\text{m}$ (compare figure E.5). This means that the numerical solution is inaccurate by less than two cells distance. Taking into account that the VOF method interpolates physical properties in the vicinity of the interface and the interface itself is not calculated as a sharp boundary but as a smeared band over multiple cells, the presented results are deemed accurate enough to justify using interFoam for capillary dominated flow problems.

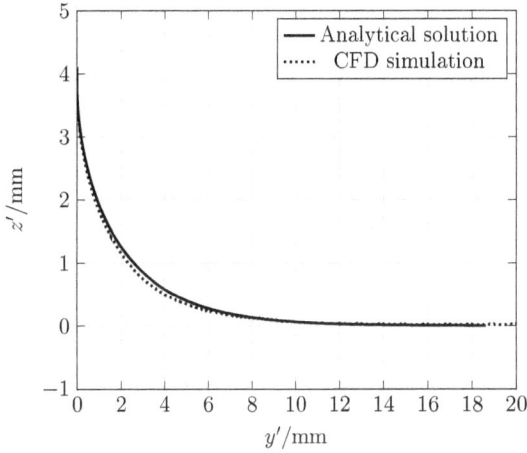

Figure E.4: Interface contour for equilibrium height of two-dimensional capillary rise at a wall. Comparison of numerical results with analytical solution.

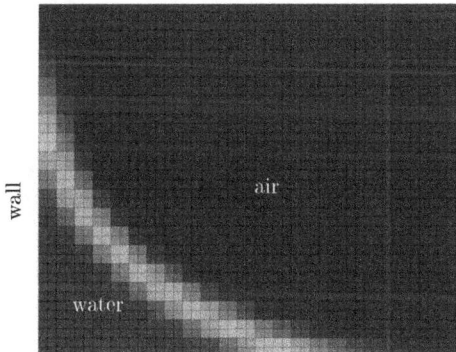

Figure E.5: Interface in vicinity of the wall. Cell outlines are coloured black, liquid is red, gas is blue. In postprocessing, the interface is determined by $\beta = 0.5$, which is coloured white here. In the VOF method, the actual interface is not calculated as a sharp boundary but smeared over multiple cells.

# Bibliography

[1] B. N. Antar and V. S. Nuotio-Antar. *Fundamentals of low gravity fluid dynamics and heat transfer.* CRC, Boca Raton, 1993.

[2] P. Ayyaswamy, I. Catton, and D. Edwards. Capillary flow in triangular grooves. *Journal of Applied Mechanics,* 41(2):332–336, 1974.

[3] D. Beattie and P. Whalley. A simple two-phase frictional pressure drop calculation method. *International Journal of Multiphase Flow,* 8(1):83–87, 1982.

[4] E. Berberović, N. P. van Hinsberg, S. Jakirlić, I. V. Roisman, and C. Tropea. Drop impact onto a liquid layer of finite thickness: Dynamics of the cavity evolution. *Physical Review E,* 79(3):036306, 2009.

[5] W. S. Bousman, J. B. McQuillen, and L. C. Witte. Gas-liquid flow patterns in microgravity: Effects of tube diameter, liquid viscosity and surface tension. *International Journal of Multiphase Flow,* 22(6):1035–1053, 1996.

[6] J. U. Brackbill, D. B. Kothe, and C. Zemach. A continuum method for modeling surface tension. *Journal of Computational Physics,* 100(2):335–354, 1992.

[7] K. A. Brakke. The Surface Evolver. *Experimental Mathematics,* 1(2):141–165, 1992.

[8] C. E. Brennen. *Fundamentals of Multiphase Flow.* Cambridge University Press, Cambridge, 2005.

[9] P. Bronowicki, P. Canfield, A. Grah, and M. E. Dreyer. Free surfaces in open capillary channels - Parallel plates. *Physics of Fluids,* 27(012106):1–21, 2015.

[10] I. N. Bronstein, K. A. Semendjajew, G. Musiol, and H. Mühlig. *Taschenbuch der Mathematik.* Verlag Harri Deutsch, 2003.

[11] R. Brown. The fundamental concepts concerning surface tension and capillarity. *Proceedings of the Physical Society,* 59(3):429, 1947.

[12] H.-J. Butt and M. Kappl. *Surface and Interfacial Forces.* Wiley-VCH, Weinheim, 2009.

[13] P. Canfield, P. Bronowicki, Y. Chen, L. Kiewidt, A. Grah, J. Klatte, R. Jenson, W. Blackmore, M. Weislogel, and M. Dreyer. The capillary channel flow experiments on the International Space Station: experiment set-up and first results. *Experiments in Fluids,* 54(5):1–14, 2013.

[14] I. Chen, R. Downing, E. Keshock, and M. Al-Sharif. Measurements and correlation of two-phase pressure drop under microgravity conditions. *Journal of Thermophysics and Heat Transfer,* 5(4):514–523, 1991.

[15] B. Choi, T. Fujii, H. Asano, and K. Sugimoto. A study of gas-liquid two-phase flow in a horizontal tube under microgravity. *Annals of the New York Academy of Sciences,* 974(1):316–327, 2002.

[16] C. Colin. Two-phase bubbly flows in microgravity: some open questions. *Microgravity Science and Technology,* 13(2):16–21, 2002.

[17] C. Colin and J. Fabre. Gas-liquid pipe flow under microgravity conditions: influence of tube diameter on flow patterns and pressure drops. *Advances in Space Research,* 16(7):137–142, 1995.

[18] C. Colin, J. Fabre, and A. E. Dukler. Gas-liquid flow at microgravity conditions - I: Dispersed bubble and slug flow. *International Journal of Multiphase Flow,* 17(4):533–544, 1991.

[19] M. Conrath, P. Canfield, P. Bronowicki, M. E. Dreyer, M. M. Weislogel, and A. Grah. Capillary channel flow experiments aboard the International Space Station. *Physical Review E,* 88(6):063009, 2013.

[20] P. Deuflhard. Newton techniques for highly nonlinear problems - theory and algorithms. unpublished, 1996.

[21] M. E. Dreyer. Strömungen mit freien Oberflächen. University lecture, 2013.

[22] M. E. Dreyer, U. Rosendahl, and H. J. Rath. Experimental investigation on flow rate limitations in open capillary flow. In *34th AIAA/ASME/SAE/ASEE Joint Propulsion Conference*. AIAA, 1998.

[23] A. Dukler, J. Fabre, J. McQuillen, and R. Vernon. Gas-liquid flow at microgravity conditions: flow patterns and their transitions. *International Journal of Multiphase Flow*, 14(4):389–400, 1988.

[24] X. Fang, H. Zhang, Y. Xu, and X. Su. Evaluation of using two-phase frictional pressure drop correlations for normal gravity to microgravity and reduced gravity. *Advances in Space Research*, 49(2):351–364, 2012.

[25] J. H. Ferziger and M. Peric. *Computational methods for fluid dynamics*. Springer, Berlin, 2012.

[26] C. Geuzaine and J.-F. Remacle. Gmsh: A 3-D finite element mesh generator with built-in pre-and post-processing facilities. *International Journal for Numerical Methods in Engineering*, 79(11):1309–1331, 2009.

[27] A. Grah and M. E. Dreyer. Dynamic stability analysis for capillary channel flow: One-dimensional and three-dimensional computations and the equivalent steady state technique. *Physics of Fluids*, 22(1):1–11, 2010.

[28] A. Grah, D. Haake, U. Rosendahl, J. Klatte, and M. E. Dreyer. Stability limits of unsteady open capillary channel flow. *Journal of Fluid Mechanics*, 600:271–289, 2008.

[29] A. Grah, P. J. Canfield, P. M. Bronowicki, M. E. Dreyer, Y. Chen, and M. M. Weislogel. Transient capillary channel flow stability. *Microgravity Science and Technology*, 26(6):385–396, 2014.

[30] D. Haake, J. Klatte, A. Grah, and M. E. Dreyer. Flow rate limitation of steady convective dominated open capillary channel flows through a groove. *Microgravity Science and Technology*, 22(2):129–138, 2010.

[31] D. E. Jaekle. Propellant management device conceptual design and analysis: Vanes. pages 1–13, 1991.

[32] S. S. Jayawardena, V. Balakotaiah, and L. C. Witte. Flow pattern transition maps for microgravity two-phase flows. *AIChE Journal*, 43(6):1637–1640, 1997.

[33] J. Klatte. *Capillary Flow and Collapse in Wedge-Shaped Channels*. PhD thesis, Universitaet Bremen, 2011.

[34] J. Klatte, D. Haake, M. M. Weislogel, and M. E. Dreyer. A fast numerical procedure for steady capillary flow in open channels. *Acta Mechanica*, 201 (1-4):269–276, 2008.

[35] M. Knorrenschild. *Numerische Mathematik*. Fachbuchverlag Leipzig im Carl-Hanser-Verlag, Munich, 2013.

[36] E. Laurien and H. Oertel. *Numerische Strömungsmechanik: Grundgleichungen und Modelle - Lösungsmethoden - Qualität und Genauigkeit*. Vieweg und Teubner, Wiesbaden, 2009.

[37] E. W. Lemmon, A. P. Peskin, M. O. McLinden, and D. Friend. NIST Thermodynamic and Transport Properties of Pure Fluids - NIST Pure Fluids. Users Guide Version 5.0, U. S. Department of Commerce, Gaithersburg, Maryland, 2000.

[38] T. Maric, J. Hopken, and K. Mooney. *The OpenFOAM Technology Primer*. Sourceflux, 2014.

[39] J. Melin, W. van der Wijngaart, and G. Stemme. Behaviour and design considerations for continuous flow closed-open-closed liquid microchannels. *Lab on a Chip*, 5:682–686, 2005.

[40] J. Meseguer, A. Sanz-Andrés, I. Pérez-Grande, S. Pindado, S. Franchini, and G. Alonso. Surface tension and microgravity. *European Journal of Physics*, 35 (5):055010, 2014.

[41] U. Nowak and L. Weimann. A family of newton codes for systems of highly nonlinear equations. Technical report, Konrad-Zuse-Zentrum fuer Informationstechnik, Berlin, 1991.

[42] L. Papula. *Mathematische Formelsammlung für Ingenieure und Naturwissenschaftler*. Vieweg und Teubner, Wiesbaden, 10. edition, 2009.

[43] T. Pedley and P. Carpenter. *Flow in Collapsible Tubes and Past Other Highly Compliant Boundaries*. Kluwer, Dordrecht, 2003.

[44] U. Rosendahl and M. E. Dreyer. Design and performance of an experiment for the investigation of open capillary channel flows. *Experiments in Fluids*, 42(5): 683–696, 2007.

[45] U. Rosendahl, A. Ohlhoff, M. E. Dreyer, and H. J. Rath. Investigation of forced liquid flows in open capillary channels. *Microgravity Science and Technology*, 13(4):53–59, 2002.

[46] U. Rosendahl, A. Ohlhoff, and M. E. Dreyer. Choked flows in open capillary channels: theory, experiment and computations. *Journal of Fluid Mechanics*, 518:187–214, 2004.

[47] U. Rosendahl, A. Grah, and M. E. Dreyer. Convective dominated flows in open capillary channels. *Physics of Fluids*, 22(052102):1–13, 2010.

[48] H. Rusche. *Computational fluid dynamics of dispersed two-phase flows at high phase fractions*. PhD thesis, Imperial College London (University of London), 2003.

[49] A. Salim, C. Colin, and M. Dreyer. Experimental investigation of a bubbly two-phase flow in an open capillary channel under microgravity conditions. *Microgravity Science and Technology*, 22(1):87–96, 2010.

[50] A. Salim, C. Colin, A. Grah, and M. E. Dreyer. Laminar bubbly flow in an open capillary channel in microgravity. *International Journal of Multiphase Flow*, 36 (9):707–719, 2010.

[51] R. Shah and A. L. London. *Laminar flow forced convection in ducts*. Academic Press, New York, 1978.

[52] R. K. Shah. A correlation for laminar hydrodynamic entry length solutions for circular and noncircular ducts. *Journal of Fluids Engineering*, 100:177–179, 1978.

[53] A. H. Shapiro. Steady flow in collapsible tubes. *Journal of Biomechanical Engineering*, 99(3):126–147, 1977.

[54] A. H. Shapiro, R. Siegel, and S. J. Kline. Friction factor in the laminar entry region of a smooth tube. In *Proceedings of the Second U. S. National Congress of Applied Mechanics*, pages 733–741, New York, 1954. American Society of Mechanical Engineers.

[55] M. Sophocleous. Understanding and explaining surface tension and capillarity: an introduction to fundamental physics for water professionals. *Hydrogeology Journal*, 18(4):811–821, 2010.

[56] E. M. Sparrow, S. H. Lin, and T. S. Lundgren. Flow development in the hydrodynamic entrance region of tubes and ducts. *Physics of Fluids*, 7(3): 338–347, 1964.

[57] R. A. Spivey, W. A. Sheredy, and G. Flores. An overview of the Microgravity Science Glovebox (MSG) facility, and the gravity-dependent phenomena research performed in the MSG on the International Space Station (ISS). In *46th AIAA Aerospace Sciences Meeting*. AIAA, 2008.

[58] J. Spurk, T. Schobeiri, and H. Marschall. *Fluid Mechanics: Problems and Solutions*. Springer Berlin Heidelberg, 2012.

[59] M. Stange. *Dynamik von Kapillarstroemungen in Zylindrischen Rohren*. PhD thesis, Universitaet Bremen, 2004.

[60] H. Tang and L. Wrobel. Modelling the interfacial flow of two immiscible liquids in mixing processes. *International Journal of Engineering Science*, 43(15):1234–1256, 2005.

[61] S. Vasavada, X. Sun, M. Ishii, and W. Duval. Study of two-phase flows in reduced gravity using ground based experiments. *Experiments in Fluids*, 43(1):53–75, 2007.

[62] Y. Wei, X. Chen, and Y. Huang. Flow rate limitation in open wedge channel under microgravity. *Science China Physics, Mechanics and Astronomy*, 56(8):1551–1558, 2013.

[63] M. M. Weislogel, E. A. Thomas, and J. C. Graf. A novel device addressing design challenges for passive fluid phase separations aboard spacecraft. *Microgravity Science and Technology*, 21(3):257–268, 2009.

[64] M. M. Weislogel, A. P. Wollman, R. M. Jenson, J. T. Geile, J. F. Tucker, B. M. Wiles, A. L. Trattner, C. DeVoe, L. M. Sharp, P. J. Canfield, J. Klatte, and M. E. Dreyer. Capillary Channel Flow (CCF) EU2-02 on the International Space Station (ISS): An Experimental Investigation of Passive Bubble Separations in an Open Capillary Channel. Technical report, National Aeronautics and Space Administration, Springfield, 2015.

[65] H. G. Weller, G. Tabor, H. Jasak, and C. Fureby. A tensorial approach to computational continuum mechanics using object-oriented techniques. *Computers in Physics*, 12(6):620–631, 1998.

[66] F. M. White. *Fluid mechanics*. McGraw-Hill, Boston, 2001.

[67] G. Woelk, M. Dreyer, and H. J. Rath. Flow patterns in small diameter vertical non-circular channels. *International Journal of Multiphase Flow*, 26(6):1037–1061, 2000.

[68] A. C. Yunus and J. M. Cimbala. Fluid mechanics: fundamentals and applications. *International Edition, McGraw Hill Publication*, pages 185–201, 2006.

[69] B. Zhao, J. S. Moore, and D. J. Beebe. Surface-directed liquid flow inside microchannels. *Science*, 291(5506):1023–1026, 2001.

[70] J. Zhao, H. Lin, J. Xie, and W. Hu. Pressure drop of bubbly two-phase flow in a square channel at reduced gravity. *Advances in Space Research*, 29(4):681–686, 2002.

[71] L. Zhao and K. S. Rezkallah. Gas-Liquid Flow Patterns at Microgravity Conditions. *International Journal of Multiphase Flow*, 19(5):751–763, 1993.

# Selbstständigkeitserklärung

Der Verfasser erklärt, dass er die vorliegende Arbeit selbständig, ohne fremde Hilfe und ohne Benutzung anderer als der angegebenen Hilfsmittel angefertigt hat. Die aus fremden Quellen (einschließlich elektronischer Quellen) direkt oder indirekt übernommenen Gedanken sind ausnahmslos als solche kenntlich gemacht. Die Arbeit ist in gleicher oder ähnlicher Form oder auszugsweise im Rahmen einer anderen Prüfung noch nicht vorgelegt worden.

_____

Bremen, 23. Januar 2018

www.ingramcontent.com/pod-product-compliance
Lightning Source LLC
Chambersburg PA
CBHW060304220326
41598CB00027B/4229